Traditional Chinese Architecture Surveying and Mapping Series: Garden Architecture

JIANGNAN GARDENS

Compiled by School of Architecture, Southeast University
Edited by CHEN Wei, SHI Fei

China Architecture & Building Press

国家出版基金项目
NATIONAL PUBLICATION FOUNDATION

「十二五」国家重点图书出版规划项目

中国古建筑测绘大系·园林建筑

江南园林

东南大学建筑学院 编写
陈薇 是霏 主编

中国建筑工业出版社

目　录

Contents

庭
园

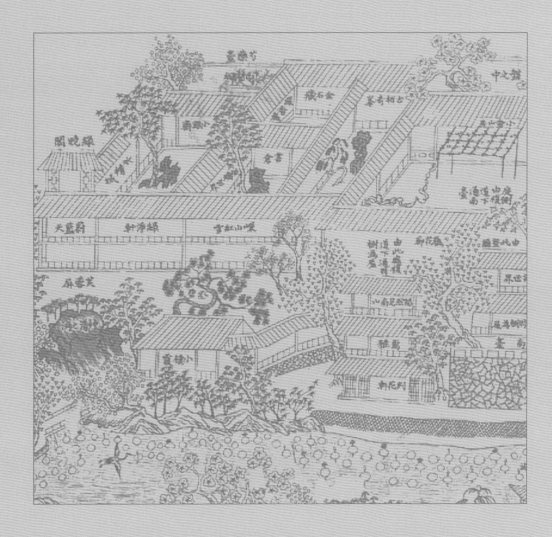

Courtyard Gardens

A courtyard garden creates a theatrical space surrounded by buildings and rockeries, dotted with flowers and shrubs. Courtyard gardens, as such, are not meant for isolated functions, but instead as an integral part of people's daily life. These gardens are typically designed in confined spaces with sophisticated scenes—a signature of Jiangnan gardens.

In courtyard gardens, one takes advantage of enclosed outdoor space by composing smaller landscape scenes using water, rockeries, and vegetations. Each small scene serves as a microcosmic depiction of nature while providing a diversified backdrop to the garden's architecture. Despite the limited space, some owners would even stage corner scenes around the courtyard periphery, creating spatial complexity and layered effects—reiterating the poetic question "how deep a courtyard could be".

The courtyard gardens holds an important place amongst classic gardens. Because of its inseparable ties to people's daily life, it remains an accessible way to evoke the natural aura in a courtyard from the past to the future. It is as practical as it is aesthetical.

庭园是由建筑物与山石花木或建筑群围合形成的空间，一般不具对外单独使用功能，而与日常生活结合紧密，小巧精致，可造景色，为江南园林一类或一种。

庭园利用围合的室外空间，凿池、置石、莳花、种树，形成小中见大的自然景色，同时使作为背景的建筑的空间产生多样变化；更有在建筑四周布置若干小院形成复合式庭院的布局，空间组织复杂，效果层层叠叠，有『庭院深深深几许』的诗意。

庭园是古典园林空间的重要组成部分，由于其与日常生活联系紧密，无论过去、现在、将来，都是人们追求自然意趣的适合方式，兼具实用价值与审美意义。

Hu Garden, Suzhou

Hu (Pot) Garden sits on the western section of the residence at No.7 Temple Lane in Suzhou. Its sophisticated water scenes brought popularity to the garden.

A moon gate marks the garden entrance which boarders a gallery leading to the southern hall and a northern flower hall . Along the gallery, a half-pavilion protrudes from the path.

The pond marks the center the garden and its scenery. The water, accompanied by two footbridges, flows beneath the flower hall and half-pavilion—enhancing the visual depth of the scenery.

The pond is circled by trees and flowers, including fern pine, crabapple, white-barked pine, bamboos, wintersweet, as well as rockeries. The compact yet diversified scenes are arranged with layered and changing views to spark contemplation.

苏州壶园

壶园在苏州庙堂巷七号，位于住宅西部，堪称以水景为主题的小庭院之上品。

入口为一圆洞门，由曲廊南与船厅相通，北与花厅相接。曲廊中段设半亭一座。

园景以水池为中心，池上架石板桥二座，池岸曲折多变，水面向花厅及半亭下延伸，增加了水源深远、余意不尽之感。

池周配以罗汉松、西府海棠、白皮松、竹丛、蜡梅等花木及湖石峰数块，景物疏密得当，恰到好处。景面层次丰富，耐人寻味。

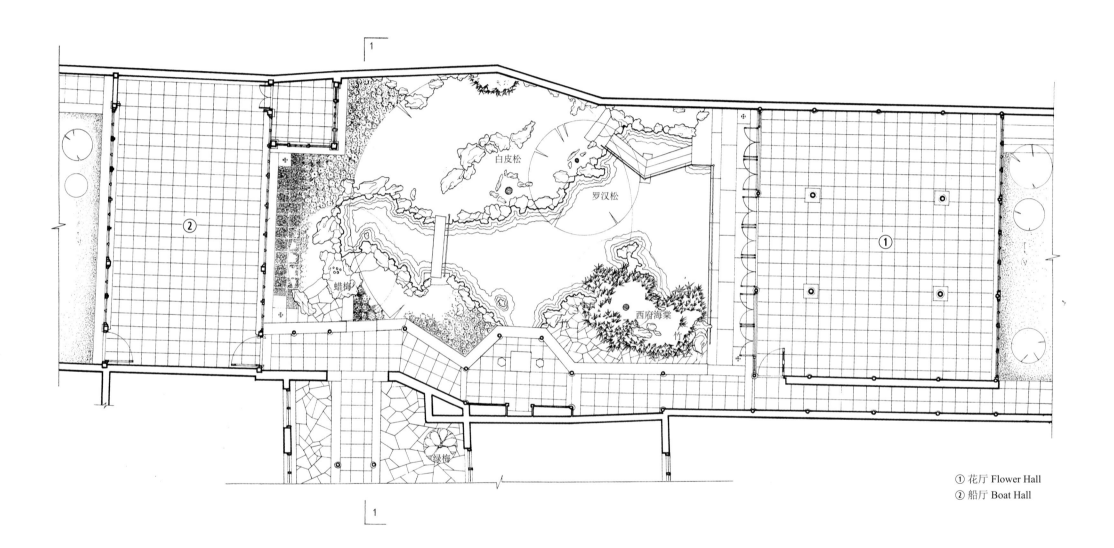

白皮松

罗汉松

蜡梅

西府海棠

绿梅

① 花厅 Flower Hall
② 船厅 Boat Hall

苏州壶园总平面图
Site plan of Hu Garden, Suzhou

N   0   1   2   3m

苏州壶园剖面图
Section of Hu Garden, Suzhou

0　　1　　2m

苏州壶园透视图
Perspective of Hu Garden, Suzhou

Wangxima Lane Garden, Suzhou

No. 7, Wangxima Lane was once the private residence of a government official named REN from Suzhou. His study garden was built during the reign of Qing emperor Guangxu.

The garden rests in the southeastern corner of the residence as a separate, walled-off area. The study sits west facing east with a gallery leading north back to the residence. To the area's south stands an elevated pavilion connected by a stepped corridor. From the study, the view is dominated by an artificial hill made of lake rockeries. One can reach the hilltop via stone steps emerging from the rockeries' northern cave, which leads to the elevated pavilion and the study thereafter.

The main garden space is planted with tall, fragrant osmanthus trees, accompanied by a flower bed dotted with *mudan* (tree peony), crape myrtle and hydrangea shrubs. A smaller flower bed is also planted on the study's back—completing an all-rounded view of the garden.

苏州王洗马巷某宅书房庭院

苏州王洗马巷七号原为任姓官僚私宅，书斋庭院建于清光绪年间。

书斋庭院位于住宅东南隅，用围墙隔成独立一区。书斋坐西向东，北面有廊通住宅，南面设小亭，以短廊相接。正对书斋前院，掇湖石假山一座，由山北边空山洞拾级而上，可通至方亭而绕回书斋。

院中有高大的桂花几株，花台植牡丹、紫薇、绣球等花木。书斋后另有一小院，设小花台，使前后有景可赏。

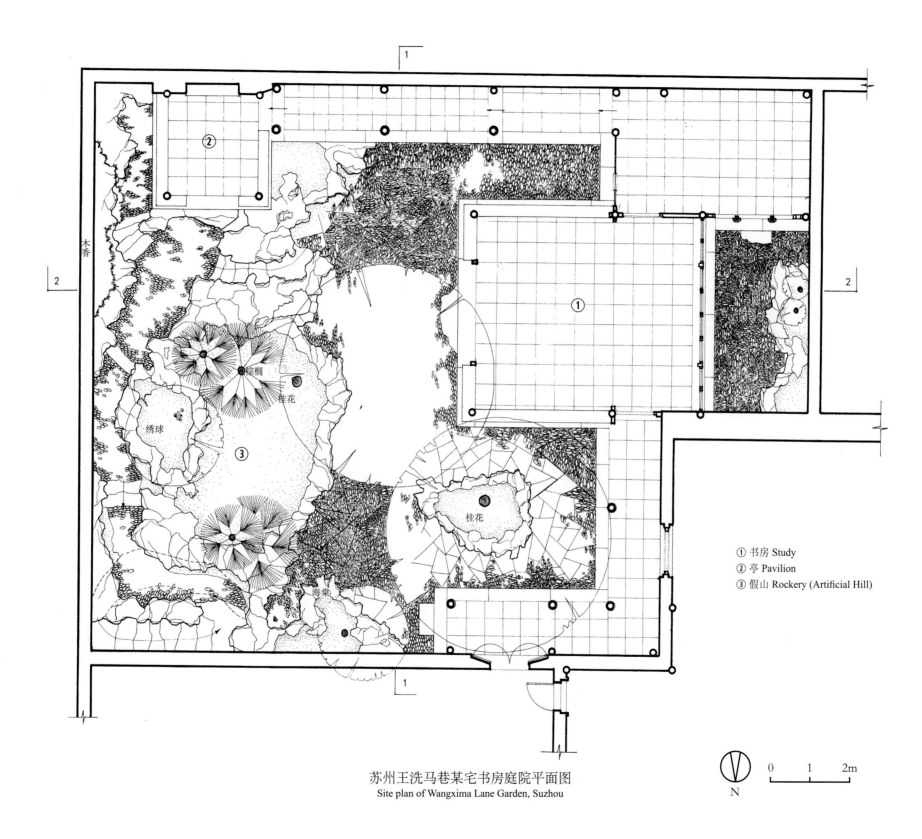

① 书房 Study
② 亭 Pavilion
③ 假山 Rockery (Artificial Hill)

苏州王洗马巷某宅书房庭院平面图
Site plan of Wangxima Lane Garden, Suzhou

0 1 2m

N

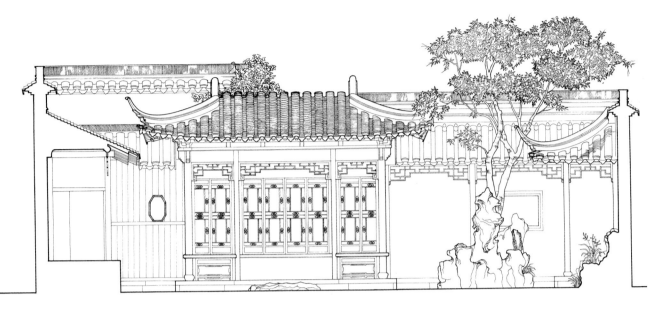

苏州王洗马巷某宅书房庭院 1-1 剖面图
Section 1-1 of Wangxima Lane Garden, Suzhou

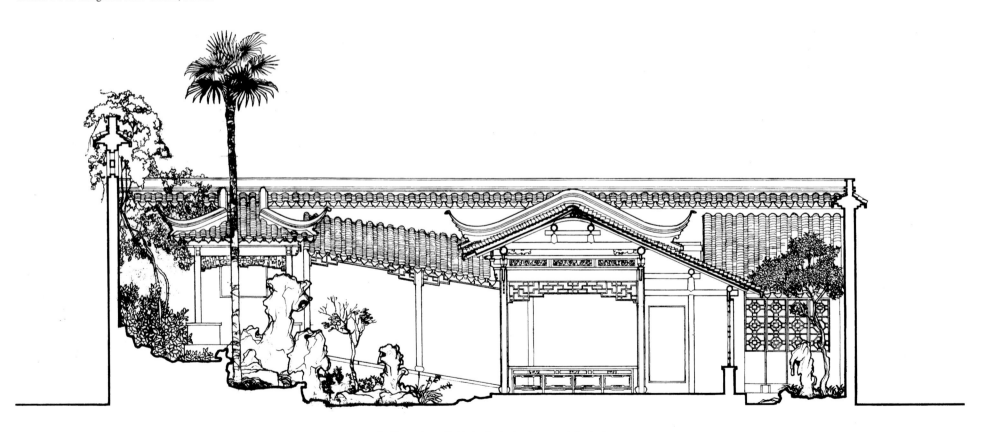

苏州王洗马巷某宅书房庭院 2-2 剖面图
Section 2-2 of Wangxima Lane Garden, Suzhou

0　　1　　2m

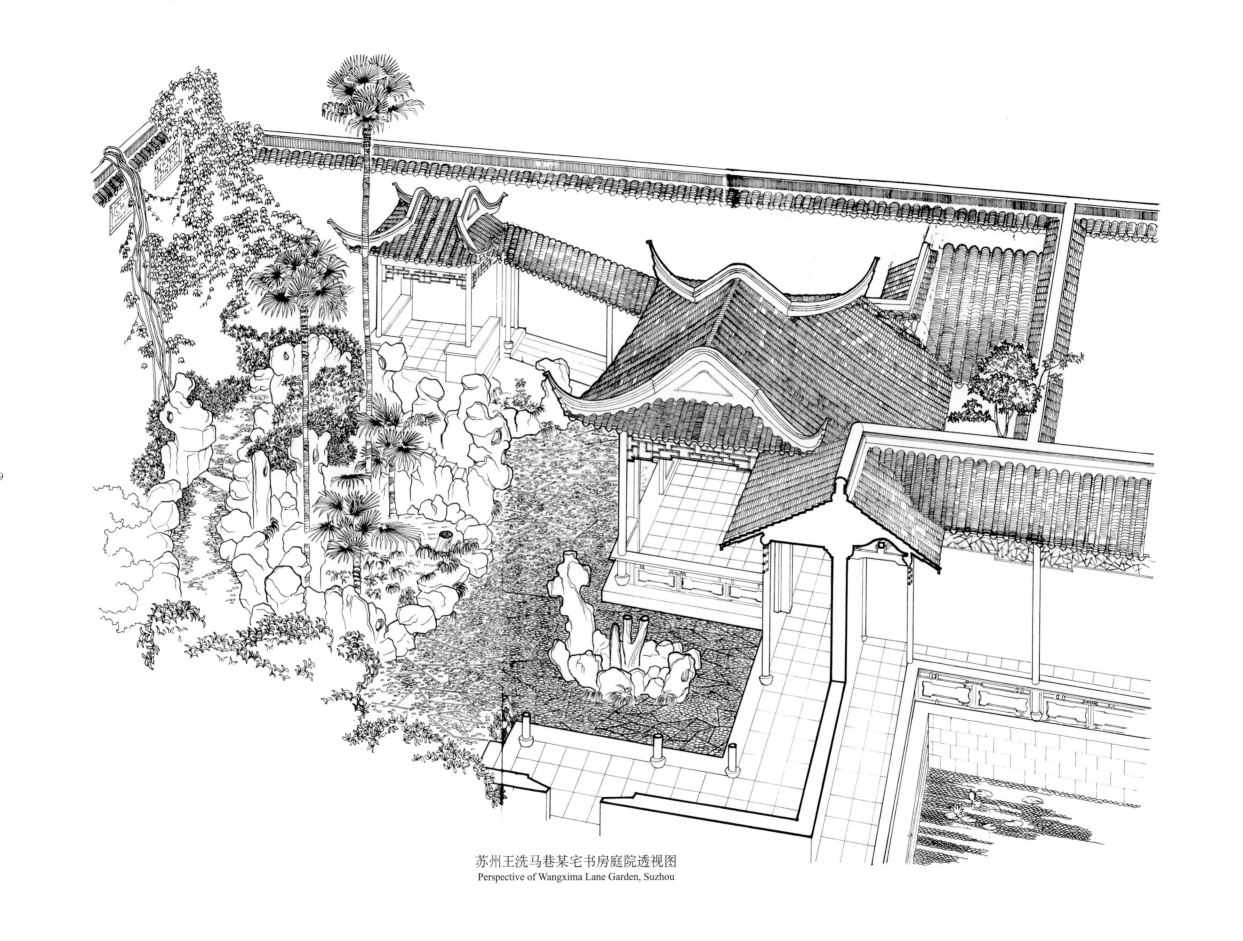

苏州王洗马巷某宅书房庭院透视图
Perspective of Wangxima Lane Garden, Suzhou

Chunxitang Courtyard, Dongting Xishan, Suzhou

The residence of Chunxitang (Bright Spring Hall) is located in Dongcai Village of Xishan in Suzhou since the reign of emperor Daoguang. The residence holds a flower hall in the southwest equipped with a front and back yard. The flower hall is 10 m wide and 9 m deep with an interior divided by a floor-to-ceiling partition. The residence hosts major activities on the northern half which faces a larger courtyard with more splendid scenery.

Both the northern and southern courtyards are dominated by flower beds, the largest of which centers a scene of flowers and scattered rockeries. The southern courtyard covers an area of 70 m sq and has a yellow rock flower bed with a boxwood tree, as well as nandina shrubs, wintersweets, and loquats. The ground is paved in an oblique-grid pattern using square slabs. The wall to the south is punctured by a hexagonal lattice window, completing a simple scene. Conversely, the northern courtyard has a larger area of 90 m sq with a central flower bed towered by three lake rockeries, the tallest reaching a steep height of 3.4 m. On the platform that hosts the flower bed stands a lush white-barked pine with a circumference up to 2 m. To top it off, the presence of a hundred-year-old *mudan* (tree peony) emits a timeless red aura across the courtyard.

苏州洞庭西山春熙堂住宅庭院

春熙堂住宅在苏州西山东蔡村，创建于清道光年间。住宅西南花厅前后各设一庭院。花厅三开间，通面阔10米，进深9米；室内以轩分隔成前后两部分，北面为主要活动场所，面向面积稍大、景物较丰的北院。

南北两院皆以花坛为主景，花坛居中，四周配花木散石。南院面积约70平方米，筑黄石花坛，植黄杨一株，并有天竹、蜡梅、枇杷等花木。地面用石板作斜方格铺地，南墙有漏窗六方，院景简洁。北院面积约90平方米，筑湖石花坛，坛上立太湖石峰3块，其中一峰高达3.4米，高而峻峭。花台上有胸围2米左右的白皮松一株，枝繁叶茂。花坛上植牡丹，百年之物，花色红艳。

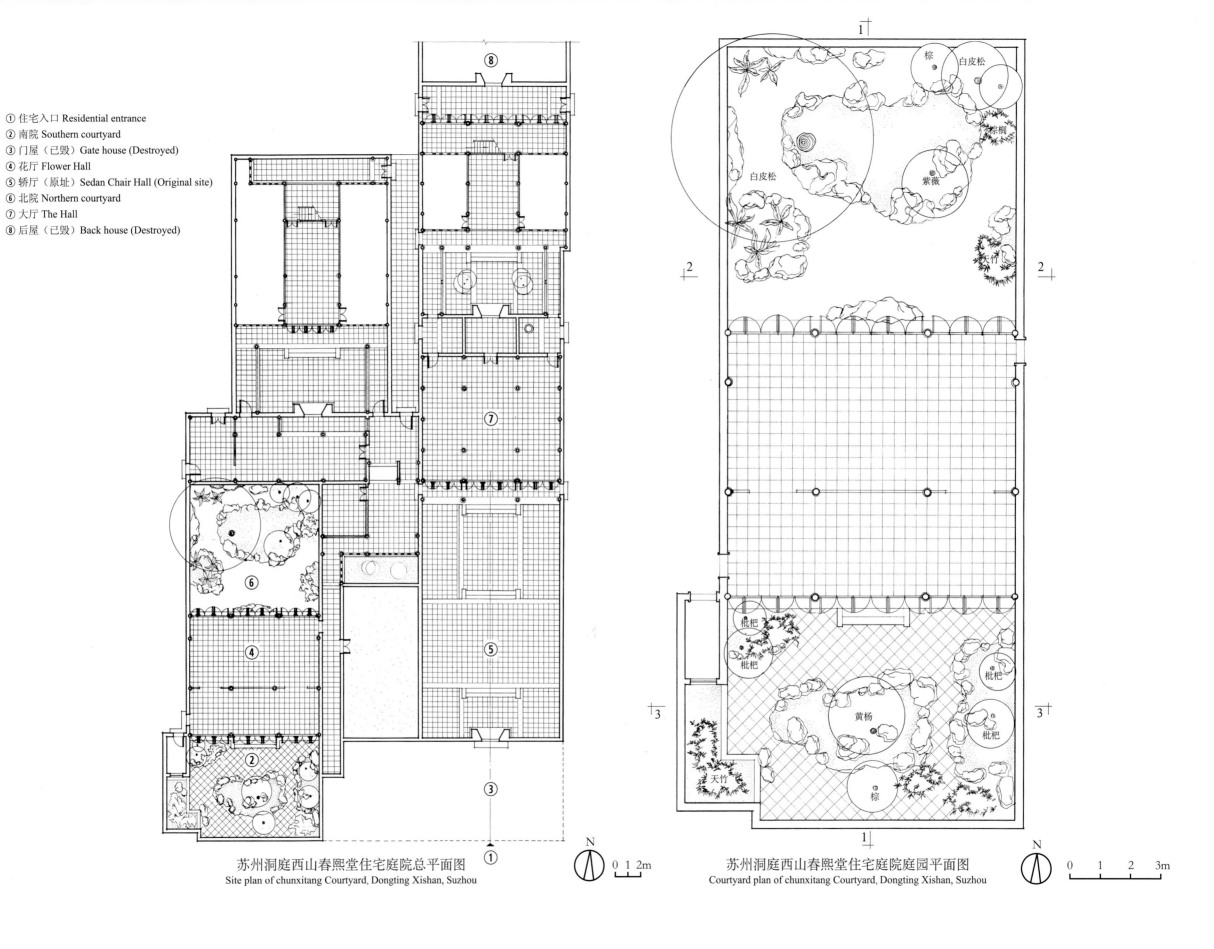

① 住宅入口 Residential entrance
② 南院 Southern courtyard
③ 门屋（已毁）Gate house (Destroyed)
④ 花厅 Flower Hall
⑤ 轿厅（原址）Sedan Chair Hall (Original site)
⑥ 北院 Northern courtyard
⑦ 大厅 The Hall
⑧ 后屋（已毁）Back house (Destroyed)

苏州洞庭西山春熙堂住宅庭院总平面图
Site plan of chunxitang Courtyard, Dongting Xishan, Suzhou

0 1 2m

苏州洞庭西山春熙堂住宅庭院庭园平面图
Courtyard plan of chunxitang Courtyard, Dongting Xishan, Suzhou

0 1 2 3m

苏州洞庭西山春熙堂住宅庭院 1-1 剖面图
Section 1-1 of Chunxitang Courtyard, Dongting Xishan, Suzhou

苏州洞庭西山春熙堂住宅庭院 2-2 剖面图
Section 2-2 of Chunxitang Courtyard, Dongting Xishan, Suzhou

苏州洞庭西山春熙堂住宅庭院 3-3 剖面图
Section 3-3 of Chunxitang Courtyard, Dongting Xishan, Suzhou

0 1 2 3m

Ai'ritang Courtyard, Dongting Xishan, Suzhou

Located in Xicai Village of Xishan in Suzhou, the courtyard of Ai'ritang (Admiring Sun Hall) was built in the late Qing dynasty. This section is in the west of the overall residence with a flower hall to the north, covering an area of 150 m sq. The three-room wide flower hall is divided into north and south sections; the larger north section is finished in a *chuanpengxuan* (boat-awning ceiling) facing the courtyard.

Yellow rockery covers the northern and western boundaries of the garden; to the east, a corridor with a sittable sill and *hejingyi* (curved-out railings) runs parallel to the wall, leaning into the courtyard. A tunnel hidden beneath the rockery is entered via flower hall's north, through which one can reach a square pavilion (destroyed, now remains its 3.6m×4.0m foundation) and a 1.1m wide pond to its south. Centering the garden is an ancient sweet olive tree with a diameter up to 40cm. Other plants include wintersweet, crape myrtle, Chinese parasol tree, bamboo and Japanese camellia. Altogether, the hall, pavilion and corridor provides an all-rounded view of the scenery and the experience of strolling through a forest.

苏州洞庭西山爱日堂庭院

爱日堂庭院在洞庭西山西蔡村，建造年代约为晚清。院落位于住宅西侧、花厅北面，面积约150平方米。花厅三开间，内部分隔成南北两部分，北部用船篷轩作天花，空间较大，面向院景。

庭院西、北两面被黄石假山所占。东面为走廊，设坐槛及鹤颈椅，可坐眺院内景色。假山有洞，由厅前拾级进洞可通至院西北隅的方亭（亭已毁，现仅存3.6米×4.0米的台基）。方亭南侧用黄石砌成直径约1.1米的小池。院内有胸径达40厘米的老桂一株，极为珍贵，其余花木为蜡梅、紫薇、梧桐、竹丛、山茶等。此院从厅、亭、廊三面可观赏假山及花木等院景，有山林意趣。

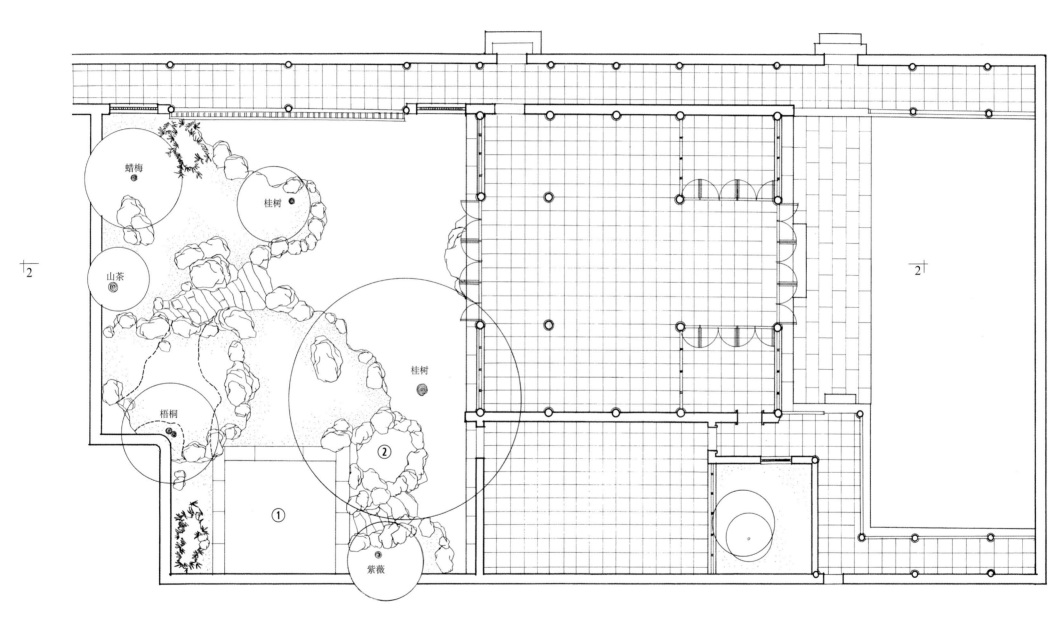

① 蜡梅
桂树
山茶
梧桐
桂树
② 紫薇

① 亭 Pavilion
② 荷花池 Lotus Pond

苏州洞庭西山爱日堂庭院总平面图
Site plan of Ai'ritang Courtyard, Dongting Xishan, Suzhou

N   0   1   2   3m

苏州洞庭西山爱日堂庭院 1-1 剖面图
Section 1-1 of Ai'ritang Courtyard, Dongting Xishan, Suzhou

苏州洞庭西山爱日堂庭院 2-2 剖面图
Section 2-2 of Ai'ritang Courtyard, Dongting Xishan, Suzhou

0　1　2　3m

Jiezhou Garden, Dongting Xishan, Suzhou

Jiezhou (Raft) Garden—located in Dongcai Village, Xishan of Suzhou—is a courtyard garden with a flower hall attached to the QIN's residence. Based on a preface from a book of the same name, the garden was built no later than 1796.

The courtyard sits to the south of the flower hall, measuring 12 m from east to west, 10 m from north to south. The flower hall shares the width of the courtyard and built with *guanyindou* (gable walls of curved profile). To the southwest of the hall, false column capitals shaped in flower baskets are suspended from the beams to convey the building's botanical function. Centering the courtyard is a stacked yellow rockery hill with hollow caves , serving as the main scenery. The rockery hill has scattered rocks and a flower bed to its west, as well as a small yellow rockery basin to its north. It is said there once stood an ancient fern pine in front of the rockeries with a diameter up to 60 cm.

苏州洞庭西山芥舟园

芥舟园在洞庭西山东蔡村，是秦氏宅中的花厅庭院。据题记『芥舟』推断，此院创建不晚于清嘉庆初年（1796年）。

庭院位于花厅之南，东西约12米，南北约10米。花厅东西与庭院同宽，山墙顶部形式采用观音兜；西南一侧作花篮厅形式，檐口下作虚柱，柱头刻花篮。庭院中以黄石叠假山一座，为院中主景。假山西侧配以散石及花坛，有洞，假山西侧配以散石及花坛，北侧用黄石筑盆池一处。假山前原有古老的罗汉松一株，胸径达60厘米以上。

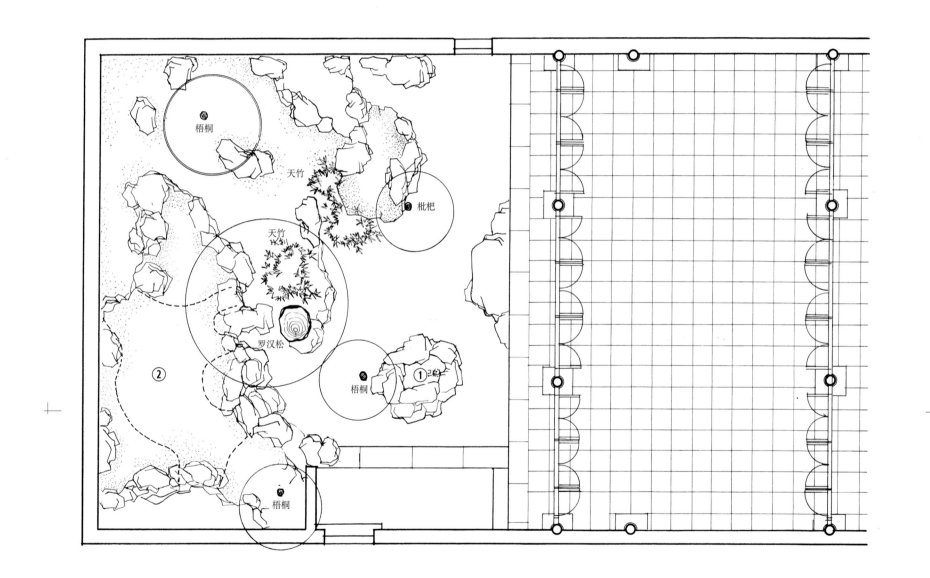

① 荷花池 Lotus pond
② 假山 Rockery (Artificial Hill)

苏州洞庭西山芥舟园总平面图
Site plan of Jiezhou Garden, Dongting Xishan, Suzhou

苏州洞庭西山芥舟园剖面图
Section of Jiezhou Garden, Dongting Xishan, Suzhou

0　1　2　3m

Nanshouxuan Courtyard, Dongting Xishan, Suzhou

Nanshouxuan (Timeless South Pavilion) Courtyard is located in Mingwan Village of Xishan, Suzhou. It is a flower hall courtyard located due east in the same residential vicinity as Lihe (Ritual Harmony) Hall. The courtyard is situated to the building's south, the ground floor of which was converted into a flower hall with modest decoration.

The courtyard is exceptionally small with a width of 6 m from east to west and a depth of 4 m from north to south. The courtyard has a flower bed stacked out of lake rockeries with a small pond encircled by plants, including bamboo, *yulan*, *mudan*, and cypress.

苏州洞庭西山南寿轩庭院

南寿轩庭院在洞庭西山明湾村，是礼和堂住宅内的花厅庭院，位于宅之东侧。庭院在楼厅之南，楼厅下层用轩作花篮厅，装修简洁。院落面积甚小，东西宽仅 6 米，南北进深约 4 米。以湖石叠花坛，凿一小池，栽植南天竹、玉兰、牡丹、柏等花木。

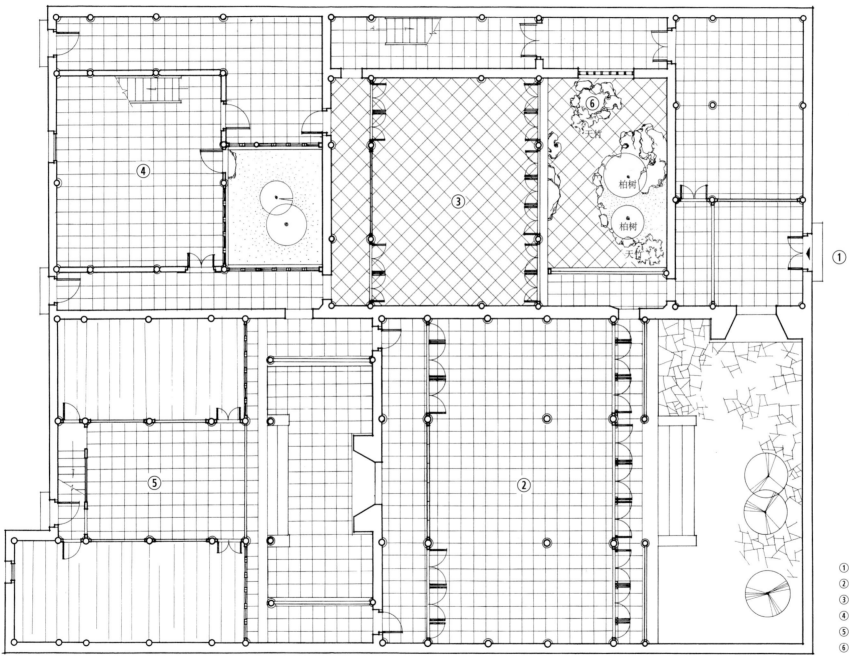

① 柏树
天树
⑥
柏树
天树

④

③

⑤

②

苏州洞庭西山南寿轩庭院总平面图
Site plan of Nanshouxuan Courtyard, Dongting Xishan, Suzhou

N

0   1   2   3m

① 宅门 Gate
② 大厅 The Hall
③ 南寿轩楼厅 South shouxuan Hall
④ 后楼 Back building
⑤ 后堂楼 Back hall building
⑥ 池 Pond

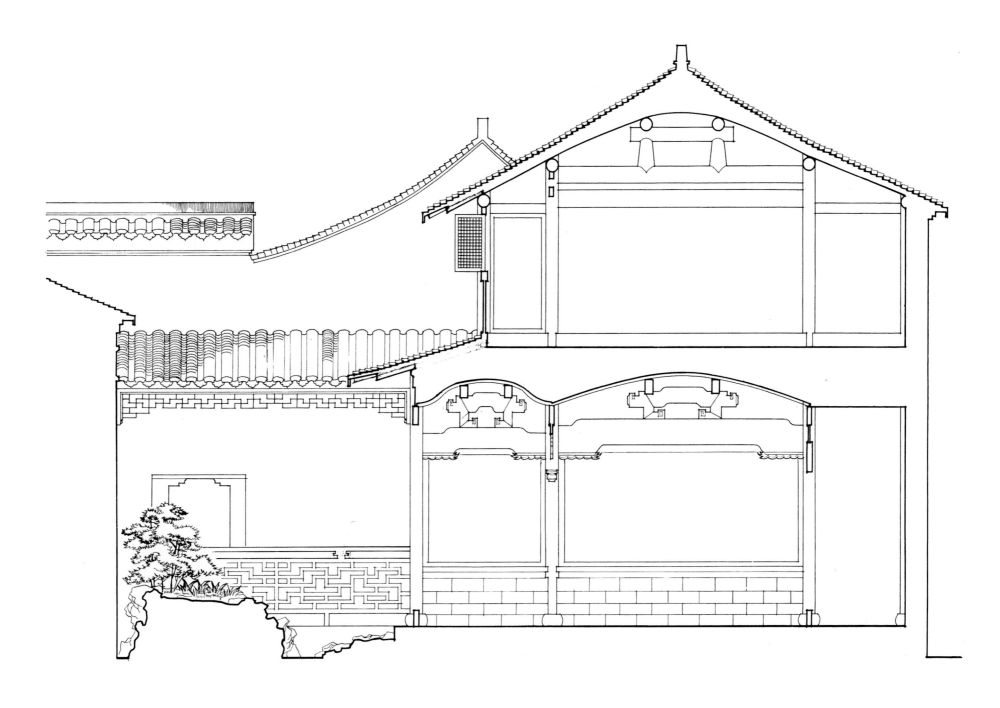

苏州洞庭西山南寿轩庭院剖面图
Section of Nanshouxuan Courtyard, Dongting Xishan, Suzhou

0　1　2　3m

Yunqilou Courtyard, Wuxi

One can find Yunqilou (Rising Cloud Building) as originally a part of the Huishan Monastery in Wuxi. The courtyard garden was first built during the reign of emperor Kangxi; the buildings today were rebuilt from the late Qing dynasty.

Situated on a mountainside, the uppermost building offers a panoramic view of Huishan Monastery; in front, a pavilion, terrace and winding gallery define the courtyard. This entire complex lies on a steep terrain dropping 4.8 m from northwest to southeast—exemplifying a typical mountain courtyard garden. Inside the courtyard, an undulating hill of yellow rockery is accompanied by fern pine, Japanese cedars, and Chinese parasol trees—emanating the ambiance of a forested mountain.

无锡惠山云起楼庭院

云起楼庭院在无锡惠山，原属惠山寺一部分，清康熙年间始建，现存建筑为清末后重建。

楼建于山腰，登楼四顾，惠山寺尽收眼底。

楼前由亭台曲廊围成庭院，地势自西北向东南逐渐下降，总高差达4.8米，是典型的山地庭院。

院内有黄石假山一座，配以罗汉松、柳杉、梧桐等树木，有山林深意。

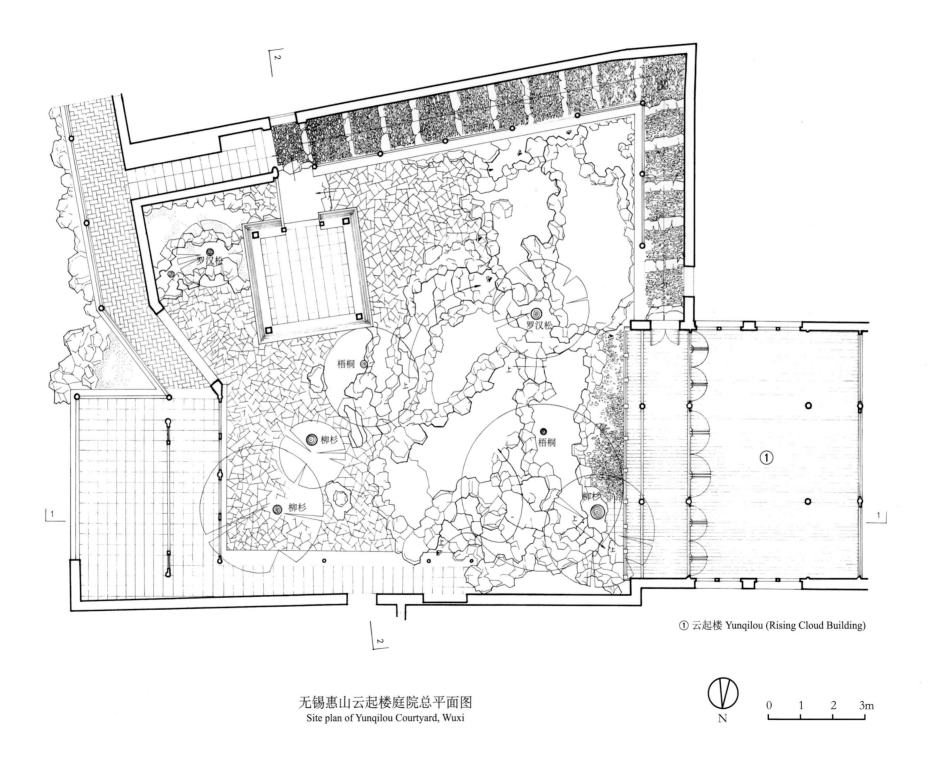

罗汉松

罗汉松

梧桐

柳杉

梧桐

柳杉

柳杉

① 云起楼 Yunqilou (Rising Cloud Building)

无锡惠山云起楼庭院总平面图
Site plan of Yunqilou Courtyard, Wuxi

N

0 1 2 3m

无锡惠山云起楼庭院 1-1 剖面图
Section 1-1 of Yunqilou Courtyard, Wuxi

无锡惠山云起楼庭院 2-2 剖面图
Section 2-2 of Yunqilou Courtyard, Wuxi

0 1 2 3m

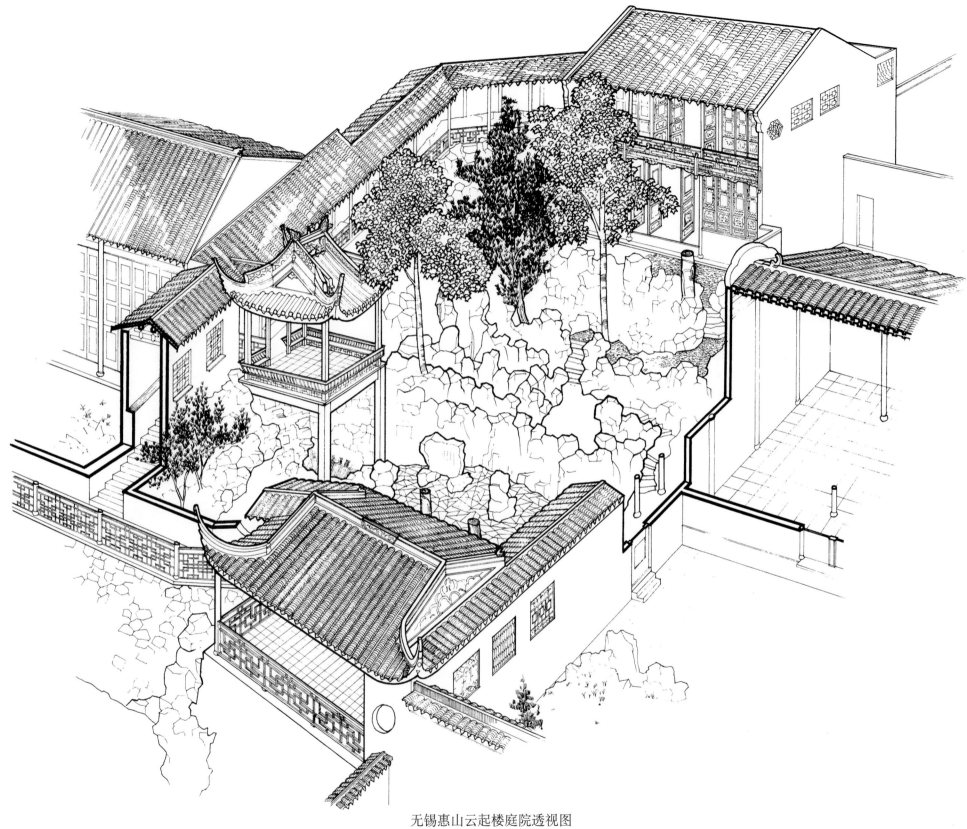

无锡惠山云起楼庭院透视图
Perspective of Yunqilou Courtyard, Wuxi

No. 6 Fengxiang Lane Courtyard, Yangzhou

No. 6 Fengxiang Lane Courtyard was built during the late Qing dynasty.

The courtyard is situated in the southern tip of the residence in front of a lower hall. The flower hall is flanked by short corridors on either side: one to its east leading to the residence, and one to its west leading to a pavilion in the southwestern corner. A rockery hill is stacked along the southern wall facing the flower hall, accompanied by a flower bed and pinnacles made from lake rockery. Adjacent to it flows a small pond above which a pavilion is perched, as if it was floating.

Though limited in space, the scenes are arranged in a neat manner. Wisteria are planted above the rockeries, while below are planted with plum tree, bamboo, Chinese privet trees and a cypress.

扬州风箱巷六号庭院

扬州新城风箱巷六号庭院，为清代晚期建筑。

庭院位于住宅南端花厅前，花厅两侧有短廊，东廊与后宅相通，西廊接至西南角半亭。面对花厅沿南墙掇假山一座，并有湖石花台及湖石峰辅之。假山旁凿一小池，池岸整齐，半亭则驾临于池上，形如水阁。

院中空间不大，布置繁简得当。山上有紫藤覆盖，山下植梅、竹，并有女贞、柏树等树木。

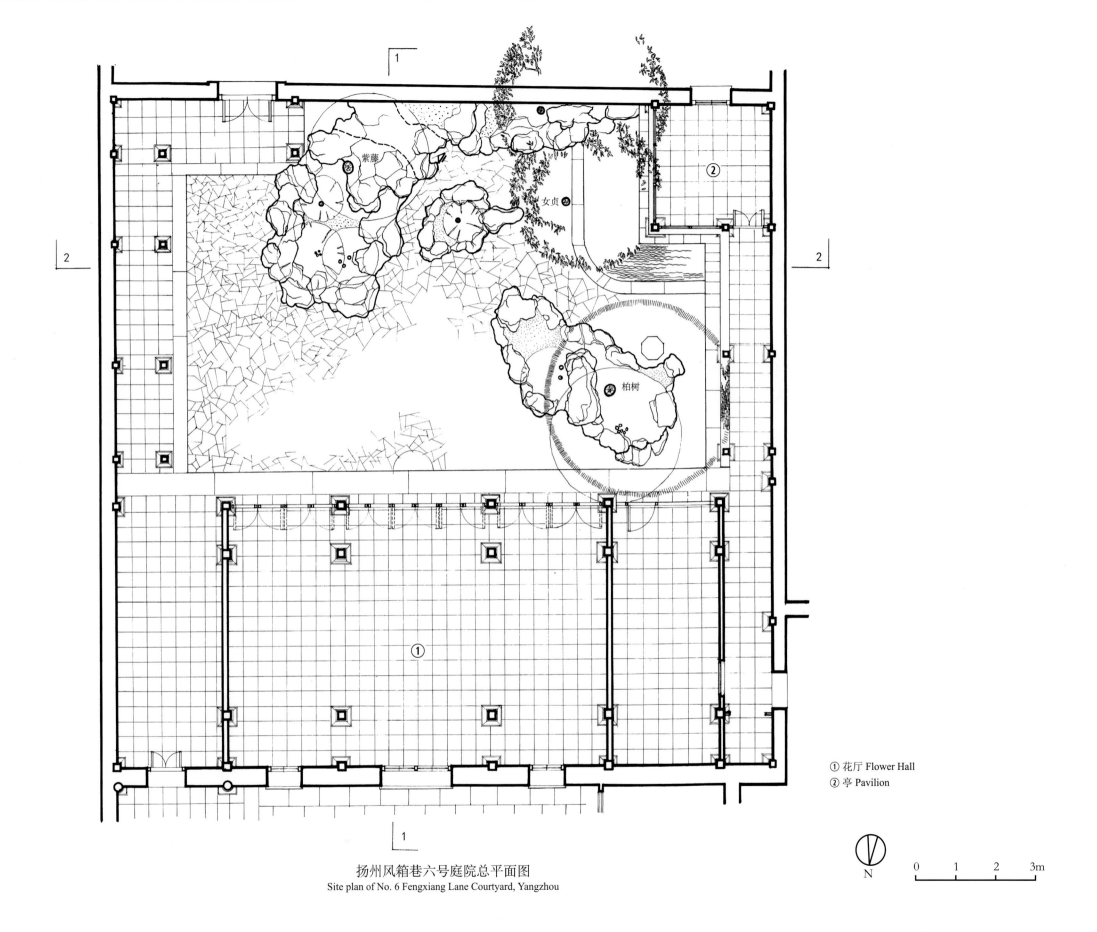

紫藤

女贞

柏树

① 花厅 Flower Hall
② 亭 Pavilion

扬州风箱巷六号庭院总平面图
Site plan of No. 6 Fengxiang Lane Courtyard, Yangzhou

N

0   1   2   3m

扬州风箱巷六号庭院 1—1 剖面图

Section 1-1 of No. 6 Fengxiang Lane Courtyard, Yangzhou

扬州风箱巷六号庭院 2—2 剖面图

Section 2-2 of No. 6 Fengxiang Lane Courtyard, Yangzhou

0　1　2　3m

No. 14 Diguan Courtyard, Yangzhou

No. 14 Diguan Courtyard is part of a residence owned by a salt merchant in Yangzhou. The residence is built with two courtyards around it. The larger front courtyard lies in the southwestern corner; the narrower backyard is named Xiaoyuan chunshen (Deep Spring in a Small Yard). The scenes staged in the courtyards are exemplary for its limited size.

The courtyard before the flower hall is separated into two by a flower wall. In the west, a corridor leads to the pavilion; from east to south, the courtyard is scattered with rockeries, flower bed, trees and a pond.

扬州地官第十四号庭院

扬州地官第十四号，为近代某盐商所建之宅，有前后两院：前院位于西南隅；后院名「小苑春深」。院景布置在小庭院中具有代表性。

前院位于花厅前，以花墙分隔为两个空间。西侧用走廊与亭子相连，东、南两侧缀以湖石峰及花台、树池。

蜡梅

桂花　白玉兰

天竹

蜡梅

扬州地官第十四号庭院总平面图
Site plan of No. 14 Diguan Courtyard, Yangzhou

N　0　1　2m

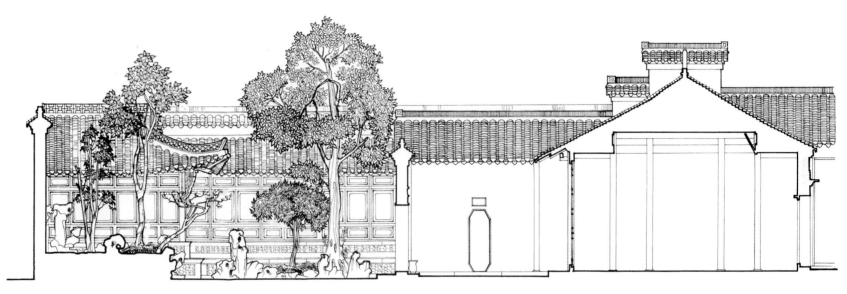

扬州地官第十四号庭院 1–1 剖面图
Section 1-1 of No. 14 Diguan Courtyard, Yangzhou

扬州地官第十四号庭院 2–2 剖面图
Section 2-2 of No. 14 Diguan Courtyard, Yangzhou

扬州地官第十四号庭院 3–3 剖面图
Section 3-3 of No. 14 Diguan Courtyard, Yangzhou

0　1　2m

Fayuan zhulin Courtyard of Zhunti'an, Nantong

The courtyard garden of Zhunti'an sits on the hillside of Langshan in Nantong, Jiangsu. The complex comprises several enclosed courtyards and was originally part of Langshan Monastery. Most of its remaining buildings are remnants repaired or rebuilt in the mid to late Qing dynasty.

The main hall, named after the Buddhist encyclopedia *Fayuan zhulin*, sits on a raised terrace overlooking the central courtyard; one can descend into the courtyard via stone steps. In the east stands Taying (Tower Shade) Hall, built during the reign of Qing emperor Daoguang. The hall was erected in large rustic timber, a style which prevailed in the early Qing dynasty. In the west sits a hall named Changle wojing (Frequent Joy, Internal Peace), which aligns with Tayin Hall and mirrors its form. In the south, a tearoom named Yizhiqi (Perched on Branch) encloses the courtyard.

Fern pine, Japanese camellia, nandina shrub, and wintersweets dot the courtyard—completing this secluded, scenic environment.

南通狼山准提庵『法苑珠林』庭院

准提庵庭院在南通狼山山腰，由若干封闭正式庭院组成，原属狼山寺院的一部分。所存建筑大部分是清中晚期遗物，后又历经重修。

主要殿堂『法苑珠林』建于高台上，居高临下，可俯览整个庭院。由石阶降至院中，东为『塔荫堂』，清道光年间建，用料粗大，尚存清初余风。西为配殿『常乐我静』，在轴线上与『塔荫堂』遥相对应，形式相仿。南为『一枝栖』茶室。

院内植罗汉松、山茶、天竹、蜡梅等花木，环境颇为清幽。

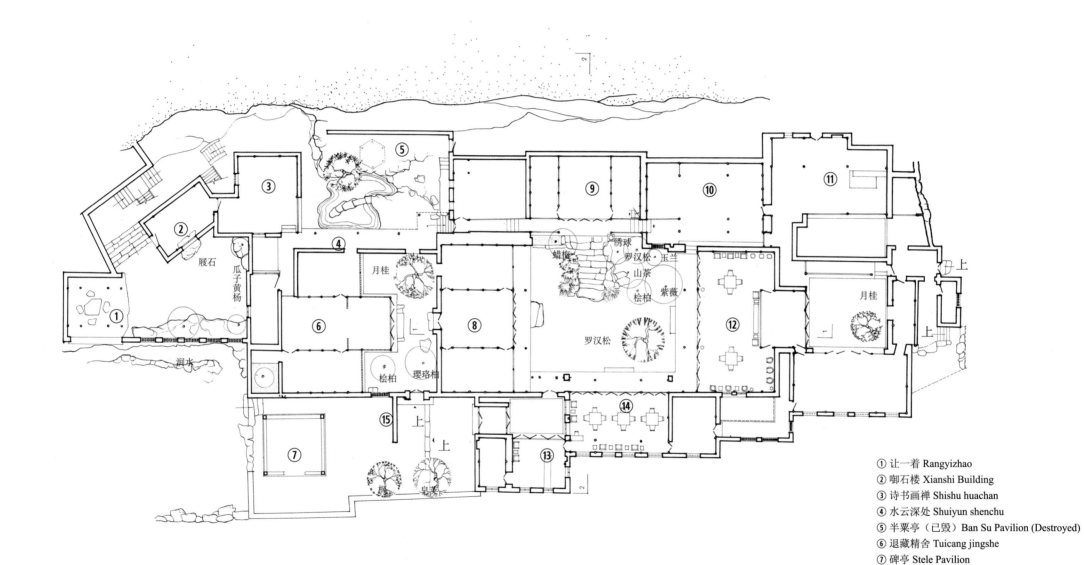

履石
瓜子黄杨
涧水
月桂
蜡梅
绣球
罗汉松
玉兰
山茶
桧柏
紫薇
罗汉松
月桂
桧柏
璎珞柏
上
上
上
上
上

南通狼山准提庵"法苑珠林"庭院总平面图
Site plan of *Fayuan zhulin* Courtyard of Zhunti'an, Nantong

① 让一着 Rangyizhao
② 啣石楼 Xianshi Building
③ 诗书画禅 Shishu huachan
④ 水云深处 Shuiyun shenchu
⑤ 半粟亭(已毁)Ban Su Pavilion (Destroyed)
⑥ 退藏精舍 Tuicang jingshe
⑦ 碑亭 Stele Pavilion
⑧ 常乐我静 Changle wojing (Frequent Joy, Internal Peace)
⑨ 法苑珠林 Fayuan zhulin
⑩ 吸江楼(在上)Xijianglou
⑪ 香积府 Xiangjifu
⑫ 塔荫堂(茶室)Tayin (Tower Shade) Hall (Tea house)
⑬ 携杖寻梅 Xiezhang xunmei
⑭ 一枝栖(茶室)Yizhiqi (Tea house)
⑮ 照壁 Screen the wall

N

0 1 2 3m

南通狼山准提庵"法苑珠林"庭院 1–1 剖面图

Section 1-1 of *Fayuan zhulin* Courtyard of Zhunti'an, Nantong

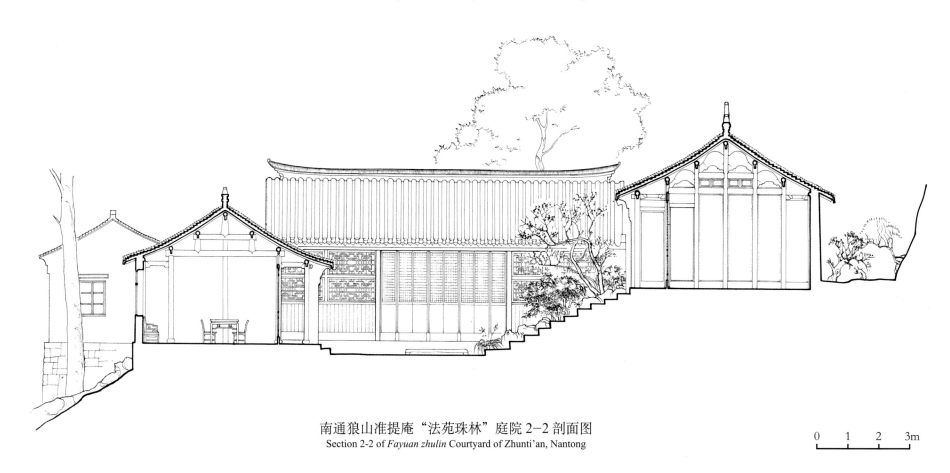

南通狼山准提庵"法苑珠林"庭院 2–2 剖面图

Section 2-2 of *Fayuan zhulin* Courtyard of Zhunti'an, Nantong

0 1 2 3m

私
园

Private Gardens

Private gardens are mainly private properties built by non-governmental owners. In the Southern Yangtze River Delta region, or Jiangnan, the literati officialdom class owned most of the private gardens. They nurtured an aesthetic taste of arts and the classic garden styles using refined gardening techniques.

Private gardens built in Jiangnan manifested a comprehensive art of spatial design. The owners applied varieties of traditional means and elements—including stacked rockeries, water scenes, buildings, flowers, trees, furnishings, furniture, poetry, paintings and carvings—to create a meaningful surrounding with rich connotations; a place to indulge themselves in leisurely peace without sacrificing their pursuit for an idealized world.

A larger private garden could be built to nest more inner gardens to create spatial layering. Assuming a courtyard garden as its unit, the private garden can be understood as a series of these units arranged sequentially and three-dimensionally. The Jiangnan private garden reached its heyday during the Ming and Qing dynasties in terms of its gardening techniques, including thematic diversity (*zhuti duoyang*), separation without obstruction (*ge'er busai*), low key before high key (*yuyang xianyi*), twisting and lingering paths (*quzhe yinghui*), appropriated scales (*chidu dedang*), rich expressions (*yuyi bujin*), and borrowing distant views (*linjie yuanjie*).

私园主要是私产、由非官方建造的园林，江南私园的主人主要是士大夫阶层，有较高的艺术审美情趣、经典的造园风格以及娴熟的造园技巧。

江南私园是运用多种中国传统艺术手段而形成的综合性的空间艺术：园主借助叠山、理水、建筑、花木、陈设、家具、诗文、绘画、雕刻等要素，营造出一种意蕴深邃、内涵丰富的生活环境，可以安适闲逸，也可以志存高远。

稍大的私园一般有园中园，空间层次丰富。若将庭园视作单元空间，那么私园空间则为序列或立体空间。江南私园到明清时期，造园手法炉火纯青，要旨有：主题多样、隔而不塞、欲扬先抑、曲折萦回、尺度得当、余意不尽、邻借远借等。

Canglangting, Suzhou

苏州沧浪亭

Canglangting (Surging Wave Pavilion) stands at No. 3 Canglang Street in Suzhou. It is the oldest classical garden in the city today. In the late Five Dynasties, the site hosted the residence of SUN Chengyou, a senior military officer of the Wu-yue kingdom. Then in mid Northern Song dynasty, SU Shunqin built the Canglang Pavilion on the waterside. The site was later expanded and became a landscape garden. The property changed hands several times, including being owned by a Buddhist monastery. Overtime, it was gradually reduced to a wasteland. However, the site underwent a major renovation during the reign of Qing emperor Kangxi: Canglang Pavilion was moved onto an earthen hill with new pavilions and galleries built around it. In addition, a stone bridge was added on the northern canal, serving as the new entrance that defines the garden's layout today. Later renovation made during the reign of emperor Daoguang and former ownership by a Buddhist monastery also opened the garden to the public, despite the subsequent deterioration. The buildings people see on site today were mostly rebuilt during the reign of emperor Tongzhi.

Covering an area up to a hectare, the garden is centered around an earthen hill. This hill stretches from west to east in an elongated form; stone pathways weaves through its lush greenery, leaving the nature untouched. The periphery is encircled by undulating galleries with stops marked by pavilions. Instead of located within, the main water feature of the garden is found outside, along its northern wall. To its west lies Mianshuixuan (Facing-water Pavilion); to its east sits the square-planned Guanyuchu (Viewing-fish Pavilion). The gallery connecting them is divided by a continuous wall punctured by perforated openings, simultaneously separates and connects the water outside to the hill within—completing a landscape of mountain and water amidst the limited site.

Canglang Pavilion is a garden with a long history; it is built on a well-situated site set amidst greenery and unadorned structures—all achieved with naturalistic techniques. Its most remarkable feat can be seen in its perforated lattice windows, of which connects internal and external scenes through beautifully crafted patterns, each unique to itself.

沧浪亭位于苏州三元坊沧浪亭街三号，是苏州古典园林中历史最久的一处。五代末年这一带曾是吴越中吴军节度使孙承佑的别墅；北宋中叶苏舜钦临水建沧浪亭，始将此处辟为园林。后历经变迁，再属佛寺，渐荒废。清康熙年间重修，移沧浪亭于土阜上，并建轩、廊等建筑，临池造石桥为入口，成为今日沧浪亭布局之基础；道光年间再修，由佛寺变祠宇，具公共性质，后毁；现存建筑多属同治年间再建时留存。

沧浪亭占地面积约一公顷，以土阜山为主，此山自西往东形态窄长，山上石径盘回，林木森郁，景色自然。建筑环山布置，随地形高低绕以走廊，配以亭榭。主水面在北面园外；西有『面水轩』（水榭），东有『观鱼处』（方亭），轩亭间连以复廊，将墙外水与墙内山连成一气。山水之间，林木葱郁。

沧浪亭历史悠久，地势优越，自然古朴，建筑朴实无华，善用漏窗疏通内外景物，且漏窗图案精美生动无一雷同，别具风格。

① 沧浪亭牌坊 Canglangting (Surging Wave Pavilion) archway
② 大门 Entrance Gate
③ 面水轩 Mianshuixuan (Facing-water Pavilion)
④ 明道堂 Mingdao Hall
⑤ 瑶华境界 Yaohua jingjie
⑥ 看山楼 Kanshan Building
⑦ 闻妙香室 Wenmiao xiangshi
⑧ 清香馆 Qingxiangguan
⑨ 五百名贤祠 Five-hundred Sages Shrine
⑩ 翠玲珑 Cuilinglong
⑪ 仰止亭 Yangzhi Pavilion
⑫ 沧浪亭 Canglangting (Surging Wave Pavilion)
⑬ 观鱼处 Guanyuchu (Viewing-fish Pavilion)
⑭ 藕花水榭 Ouhua shuixie
⑮ 御碑亭 Imperial stele Pavilion

苏州沧浪亭总平面图
Site plan of Canglangting, Suzhou

N

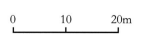
0   10   20m

苏州沧浪亭 1-1 剖面图
Section 1-1 of Canglangting, Suzhou

0　　2　　4m

苏州沧浪亭 2—2 剖面图
Section 2-2 of Canglangting, Suzhou

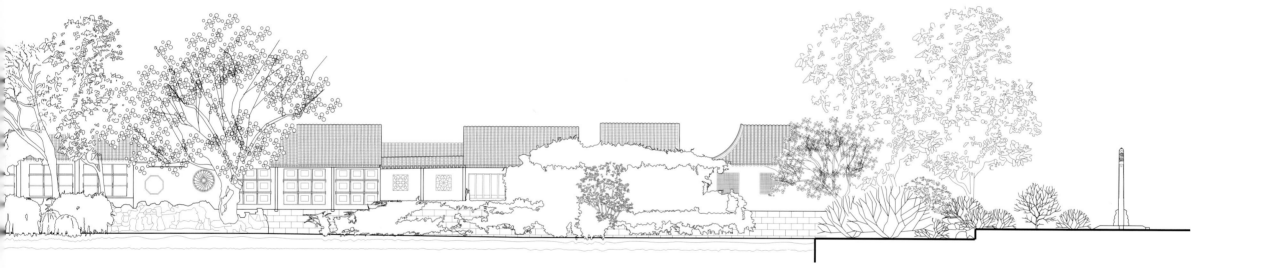

0　2　4m

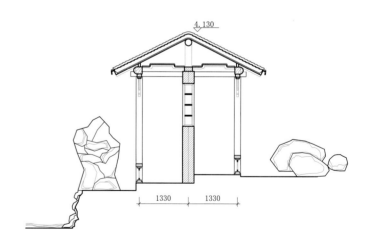

4,130

1330 1330

苏州沧浪亭复廊剖面图
Section of northern gallery of Canglangting, Suzhou

0   2   4m

苏州沧浪亭复廊内侧展开立面图
Inner elevation of northern gallery of Canglangting, Suzhou

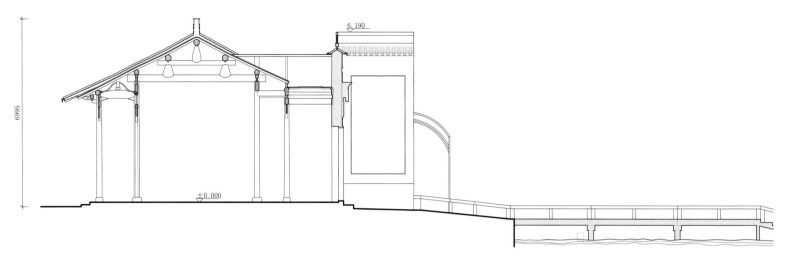

苏州沧浪亭门厅明间剖面图

Section of central-bay of entrance hall of Canglangting, Suzhou

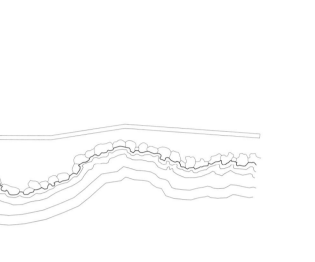

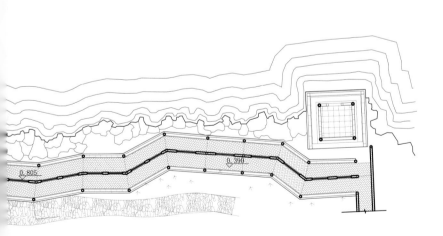

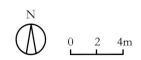

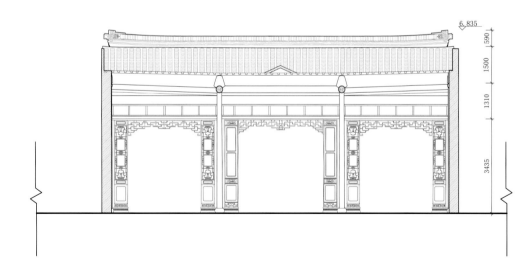

苏州沧浪亭门厅纵剖面图

Longitudinal section of entrance hall of Canglangting, Suzhou

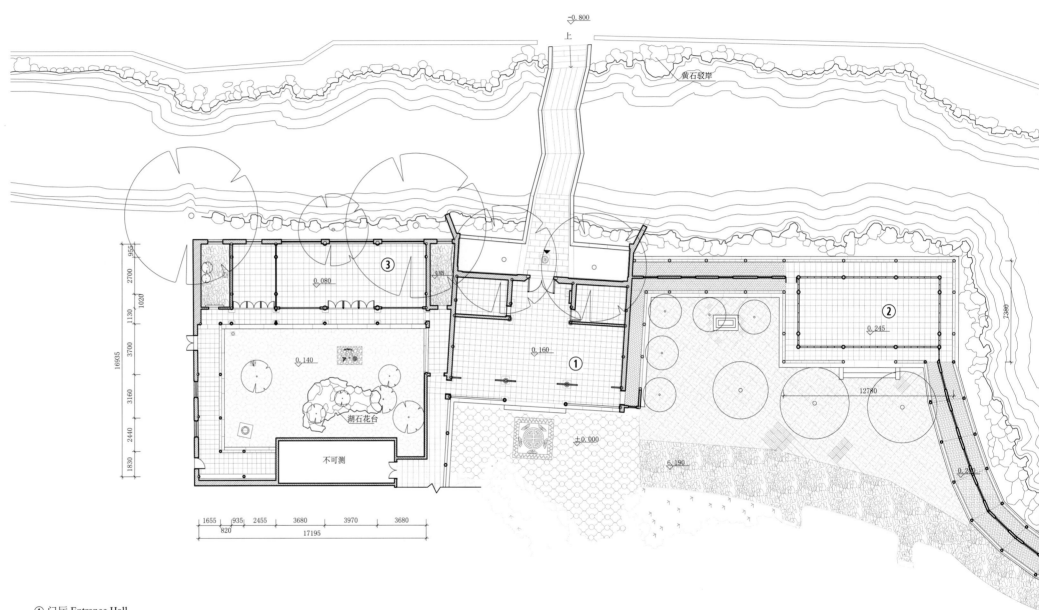

① 门厅 Entrance Hall
② 面水轩 Mianshuixuan (Facing-water Pavilion)
③ 藕花水榭 Ouhuashuixie

苏州沧浪亭入口区平面图
Entrance area Site plan of Canglangting, Suzhou

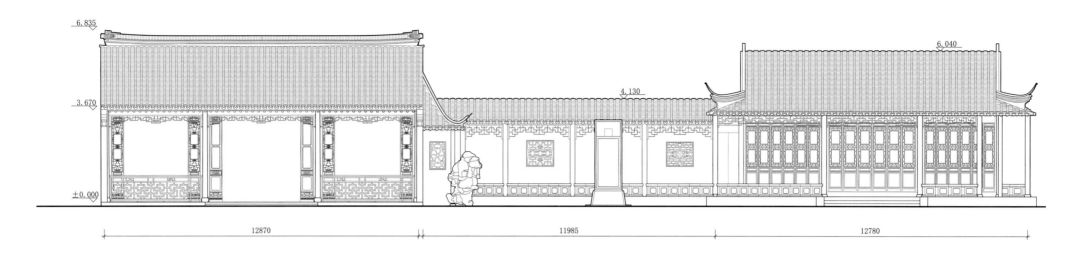

6.835

3.670

±0.000

6.040

4.130

12870　　11985　　12780

苏州沧浪亭门厅－面水轩立面图
Elevation from entrance hall to Mianshuixuan of Canglangting, Suzhou

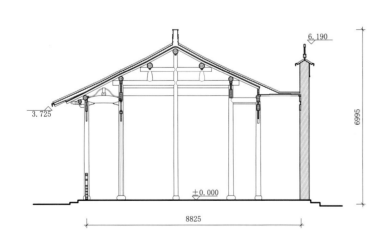

6.190

3.725

±0.000

6995

8825

苏州沧浪亭门厅次间剖面图
Section of Side-bay of entrance hall of Canglangting, Suzhou

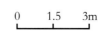

0　1.5　3m

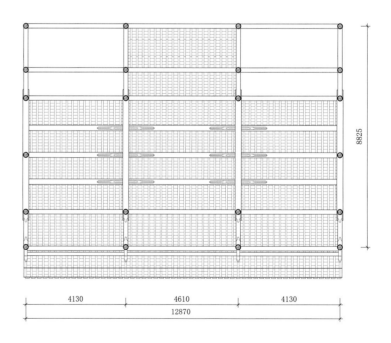

8825

4130　4610　4130
12870

苏州沧浪亭门厅仰视平面图
Reflected ceiling plan of entrance hall of Canglangting, Suzhou

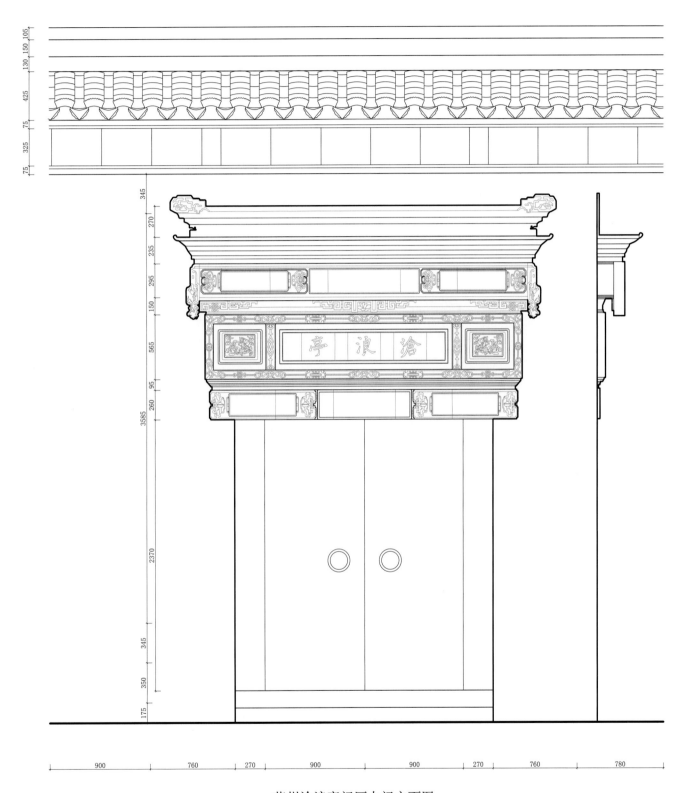

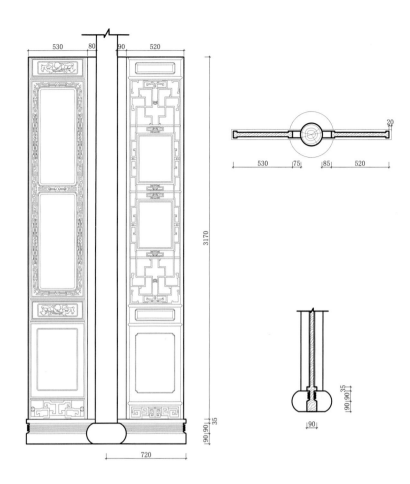

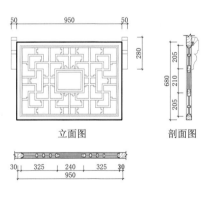

立面图

剖面图

苏州沧浪亭门厅大门立面图

Elevation of entrance gate of Canglangting, Suzhou

苏州沧浪亭藕花水榭推窗大样图

Windows in Ouhuashuixie of Canglangting, Suzhou

0   0.3   0.6m

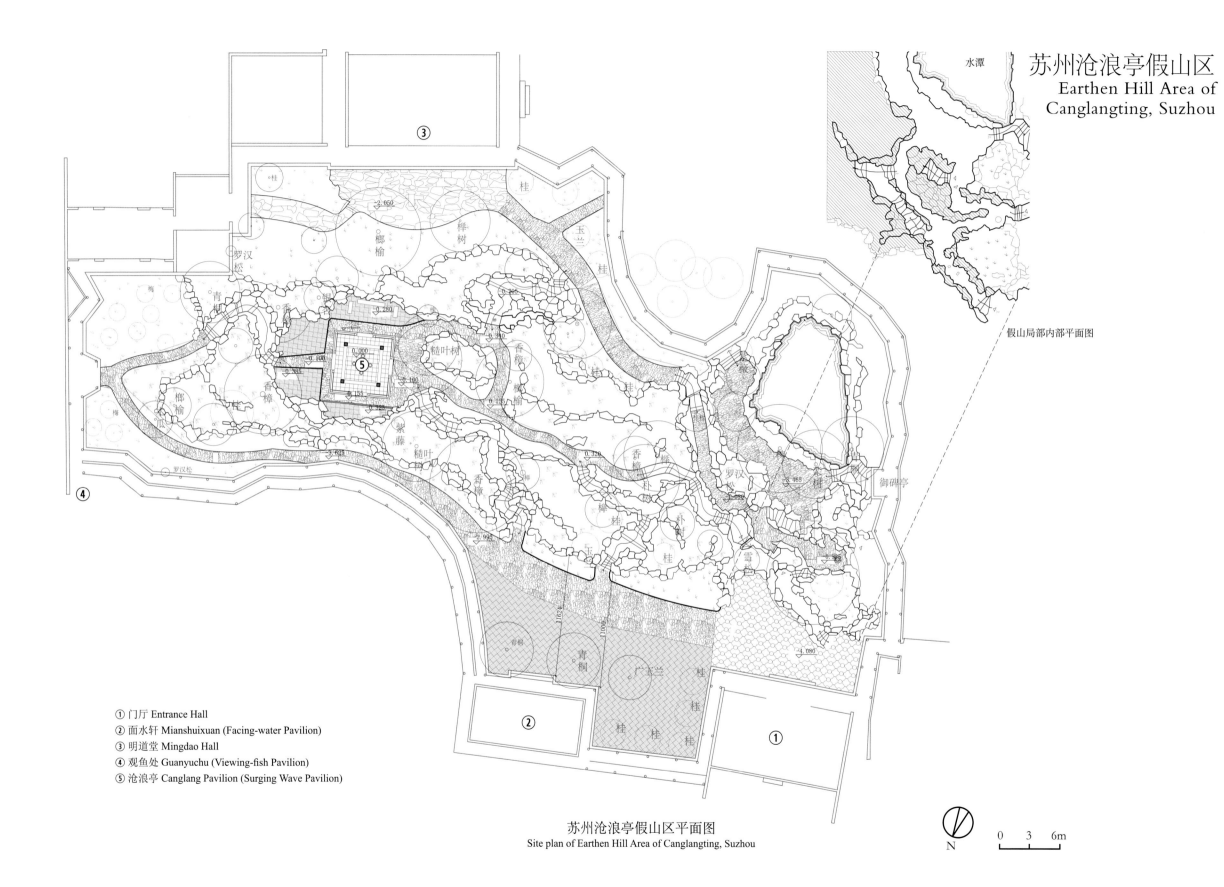

水潭

假山局部内部平面图

① 门厅 Entrance Hall
② 面水轩 Mianshuixuan (Facing-water Pavilion)
③ 明道堂 Mingdao Hall
④ 观鱼处 Guanyuchu (Viewing-fish Pavilion)
⑤ 沧浪亭 Canglang Pavilion (Surging Wave Pavilion)

苏州沧浪亭假山区平面图
Site plan of Earthen Hill Area of Canglangting, Suzhou

N

0   3   6m

明道堂

A

B

面水轩

入口门厅

C

3200 1660 3560 2155 630 700 7640

苏州沧浪亭假山区立面图 A
Elevation A of Earthen Hill Area of Canglangting, Suzhou

0 0.3 0.6m

苏州沧浪亭假山区立面图 B
Elevation B of Earthen Hill Area of Canglangting, Suzhou

3865

6510 3850

苏州沧浪亭假山区立面图 C
Elevation C of Earthen Hill Area of Canglangting, Suzhou

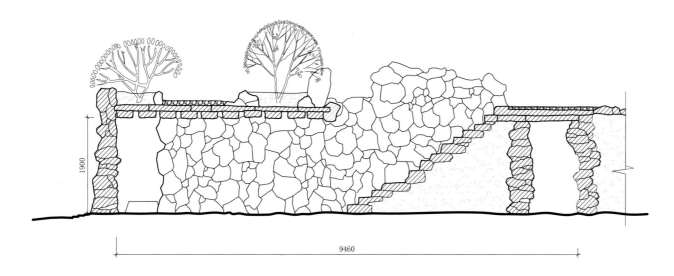

苏州沧浪亭假山局部 1-1 剖面图
Section 1-1 of earthen hill of Canglangting, Suzhou

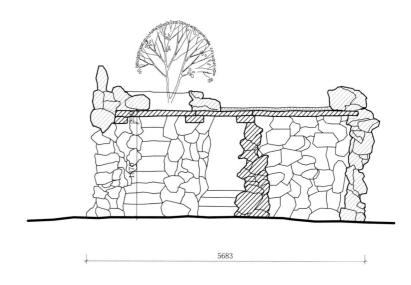

苏州沧浪亭假山局部 2-2 剖面图
Section 2-2 of earthen hill of Canglangting, Suzhou

0   1   2m

苏州沧浪亭假山洞口立面图 A
Elevation A of entrance to cave of earthen hill of Canglangting, Suzhou

苏州沧浪亭假山洞口立面图 B
Elevation B of entrance to cave of earthen hill of Canglangting, Suzhou

苏州沧浪亭假山洞口立面图 C
Elevation C of entrance to cave of earthen hill of Canglangting, Suzhou

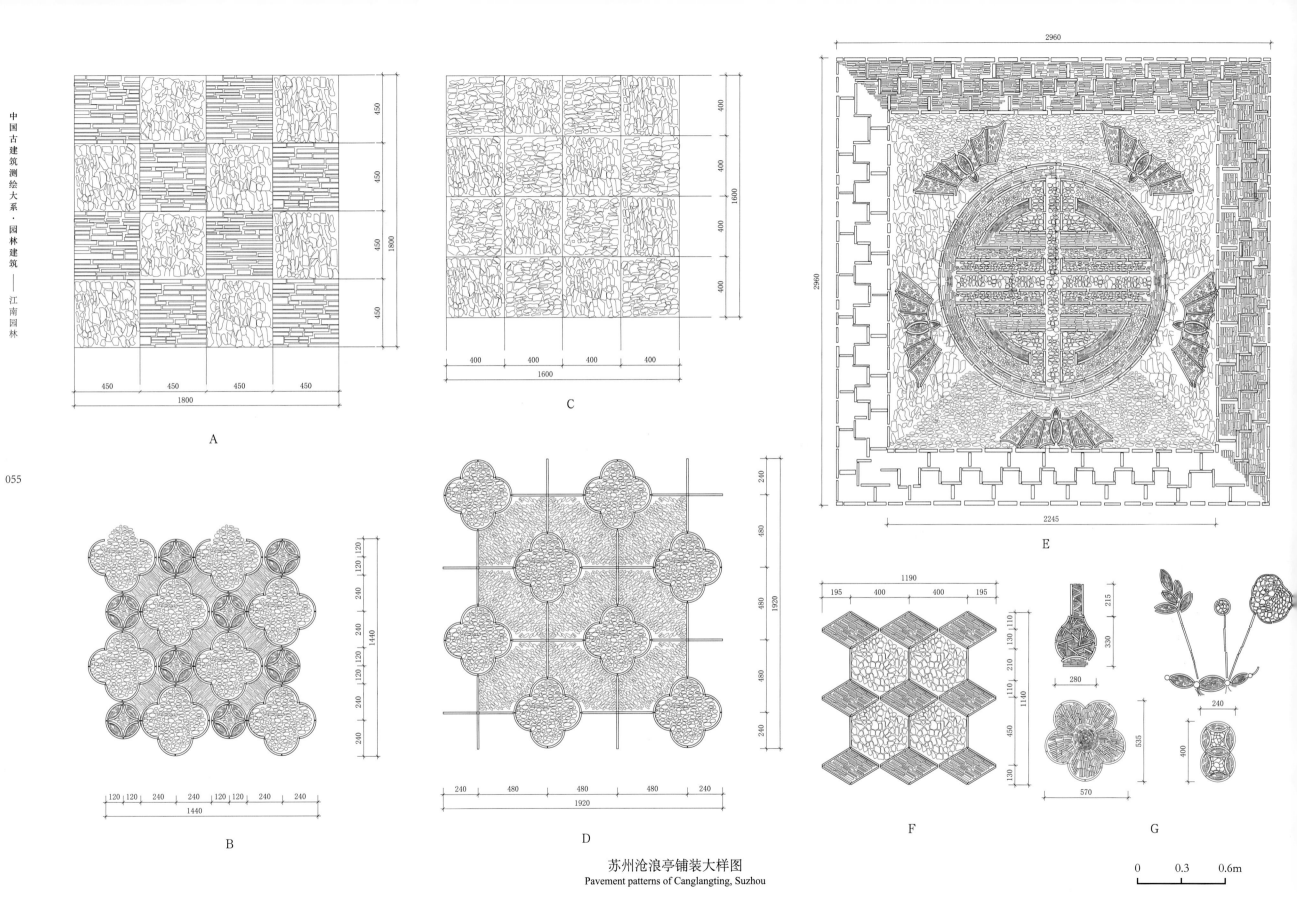

A

B

C

D

E

F

G

苏州沧浪亭铺装大样图
Pavement patterns of Canglangting, Suzhou

0   0.3   0.6m

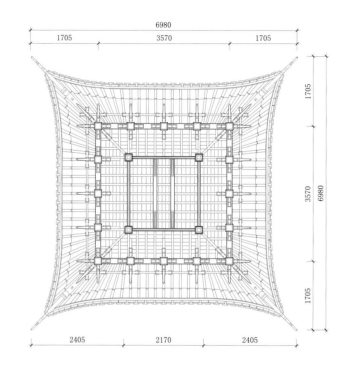

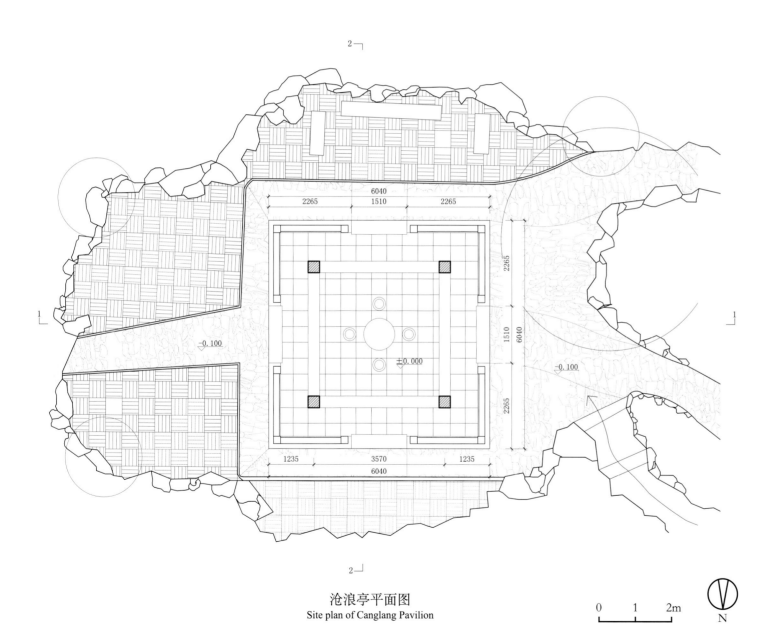

沧浪亭平面图
Site plan of Canglang Pavilion

0 1 2m

N

沧浪亭仰视平面图
Reflected ceiling plan of Canglang Pavilion

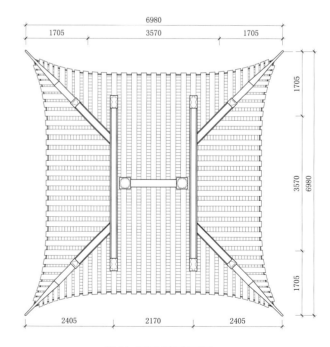

沧浪亭屋顶平面图
Roof plan of Canglang Pavilion

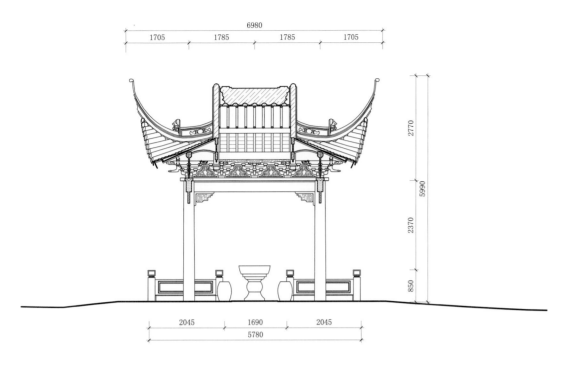

沧浪亭 1-1 剖面图

Section 1-1 of Canglang Pavilion

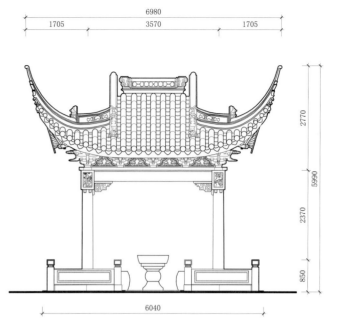

沧浪亭南立面图

South elevation of Canglang Pavilion

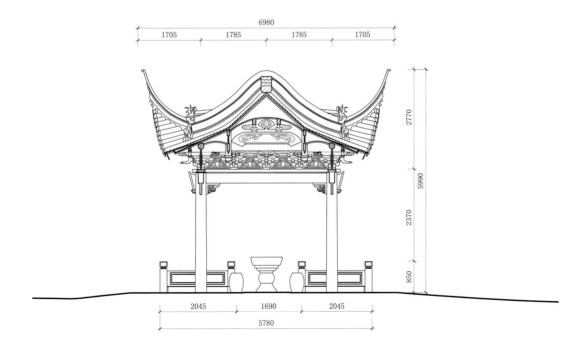

沧浪亭 2-2 剖面图

Section 2-2 of Canglang Pavilion

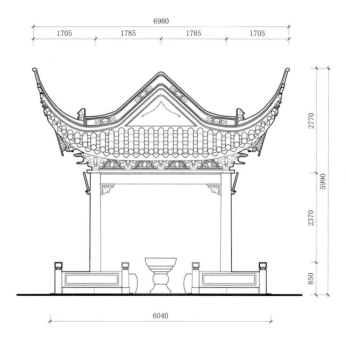

沧浪亭东立面图

East elevation of Canglang Pavilion

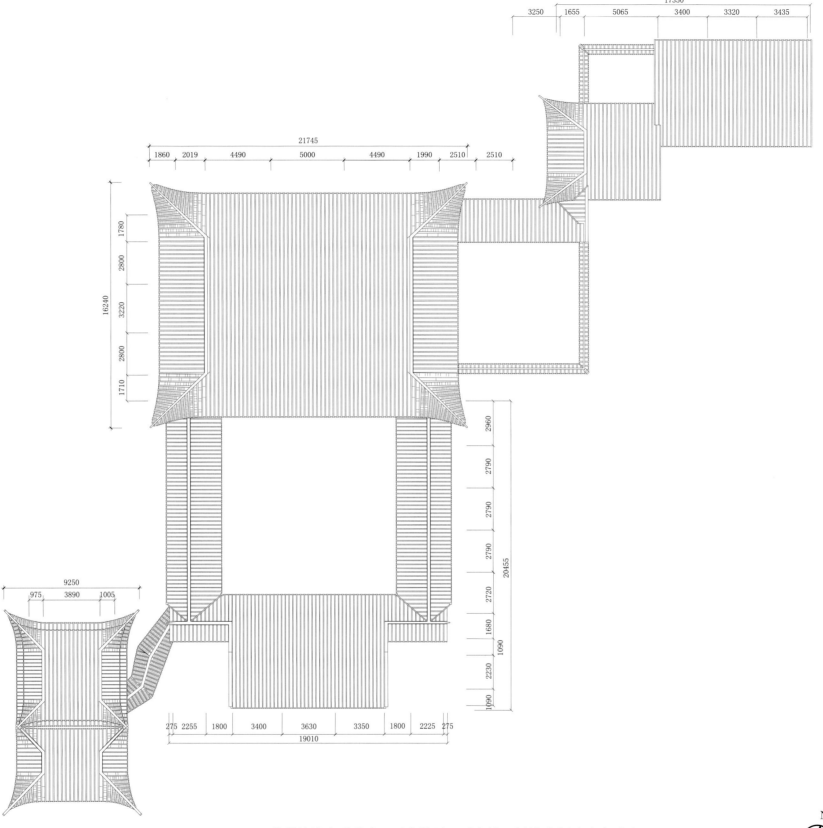

苏州沧浪亭明道堂—瑶华境界—看山楼—闻妙香室屋顶平面图
Roof plan of Mingdao Hall, Yaohua jingjie, Kanshan Building, Wenmiao xiang shi of Canglangting, Suzhou

N

0   2.5   5m

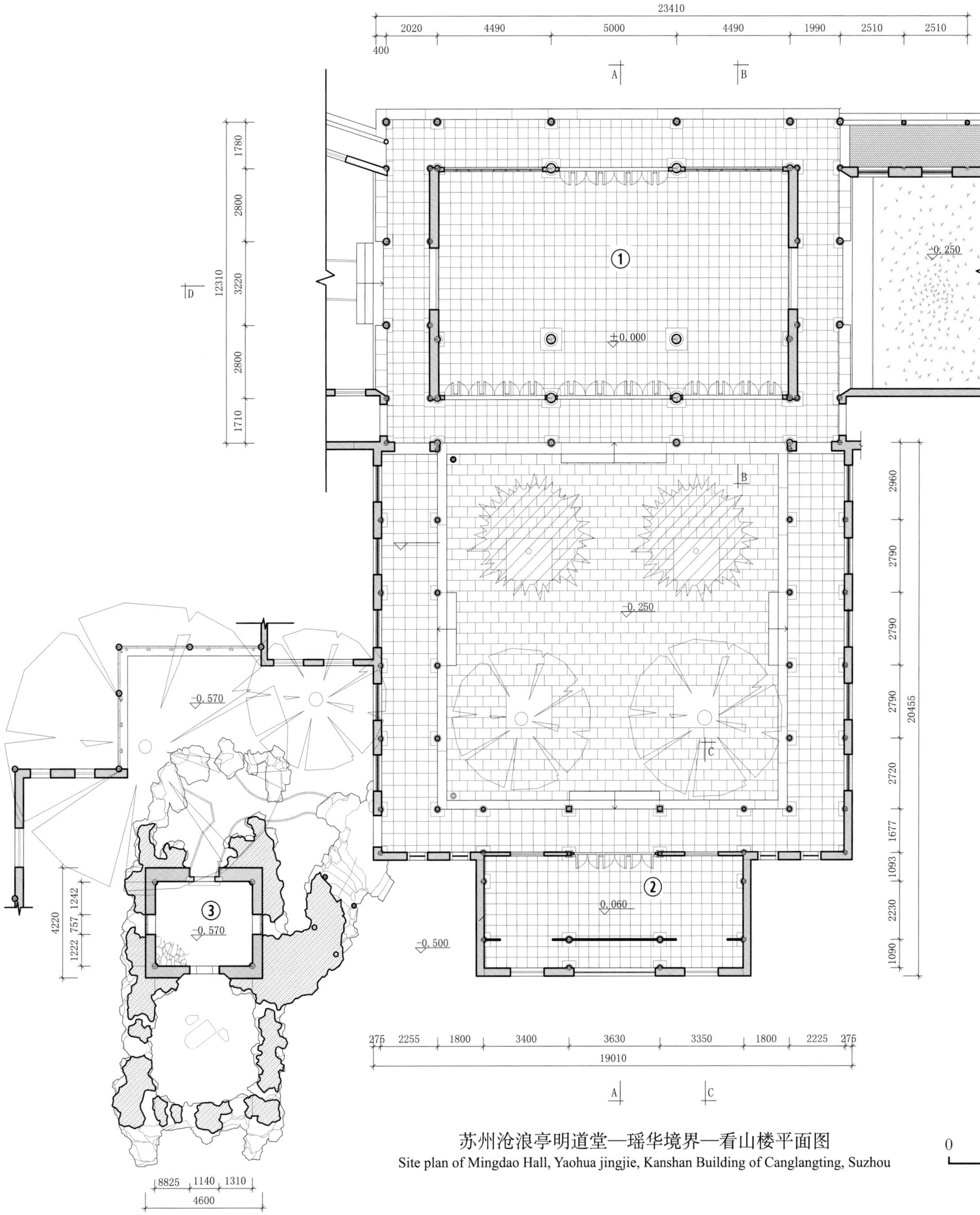

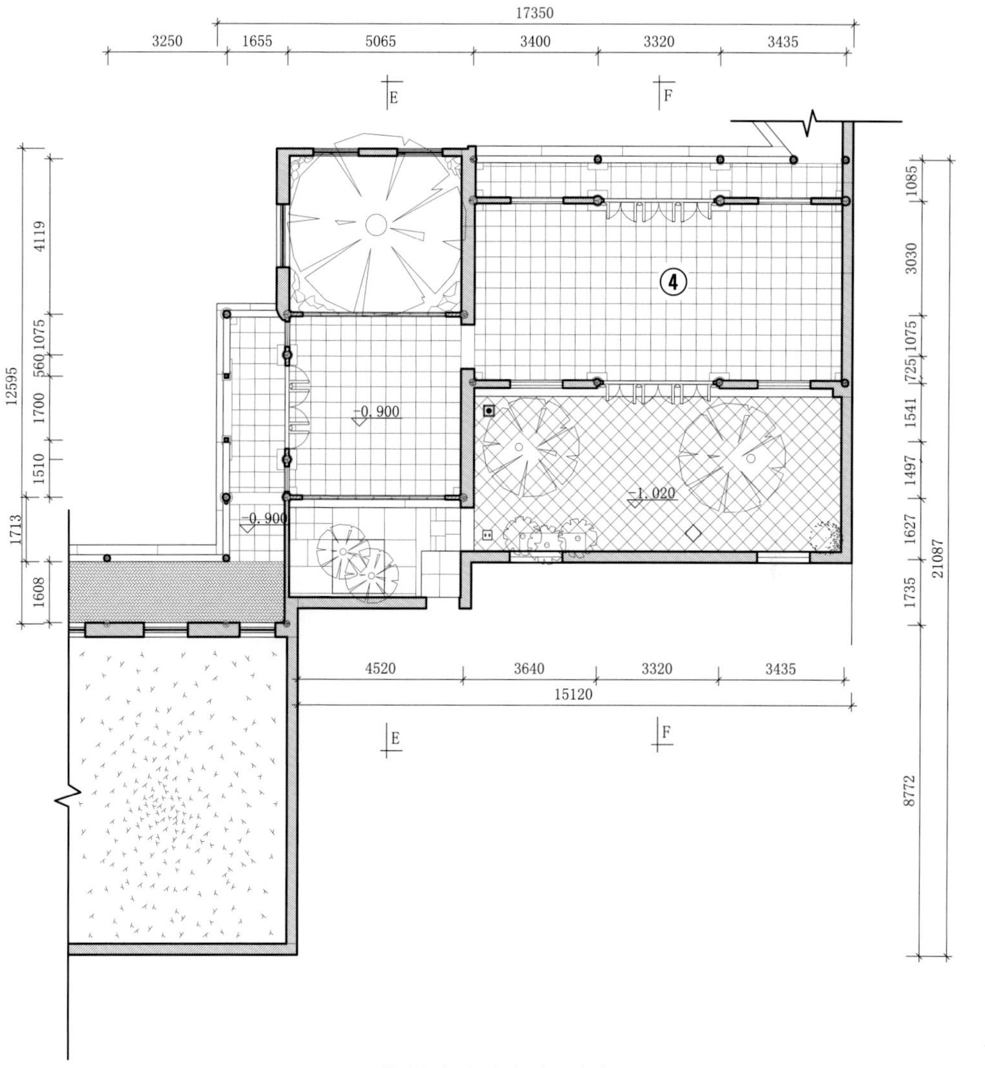

059

苏州沧浪亭闻妙香室平面图
Plan of Wenmiao xiangshi of Canglangting, Suzhou

① 明道堂 Mingdao Hall
② 瑶华境界 Yaohua jingjie
③ 看山楼 Kanshan Building
④ 闻妙香室 Wenmiao xiangshi

苏州沧浪亭明道堂—瑶华境界—看山楼平面图
Site plan of Mingdao Hall, Yaohua jingjie, Kanshan Building of Canglangting, Suzhou

0   2   4m

N

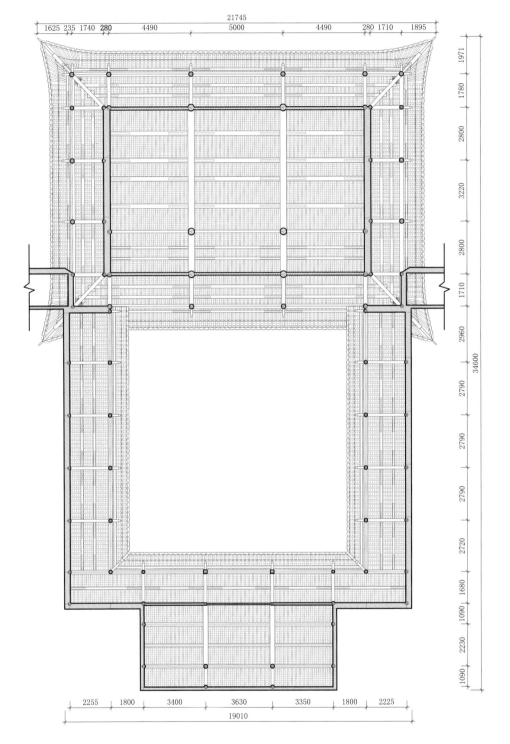

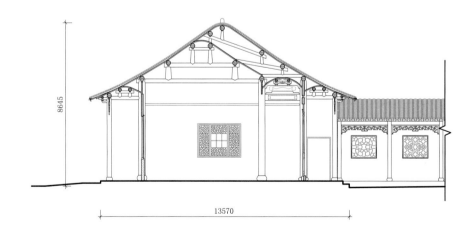

苏州沧浪亭明道堂—瑶华境界—看山楼仰视平面图
Reflected ceiling plan of Mingdao Hall, Yaohua jingjie, Kanshan Building of Canglangting, Suzhou

苏州沧浪亭明道堂—瑶华境界—看山楼 A-A 剖面图
Section A-A of Mingdao Hall, Yaohua jingjie, Kanshan Building of Canglangting, Suzhou

苏州沧浪亭明道堂—瑶华境界—看山楼 B-B 剖面图
Section B-B of Mingdao Hall, Yaohua jingjie, Kanshan Building of Canglangting, Suzhou

N

0　2　4m

中国古建筑测绘大系·园林建筑——江南园林

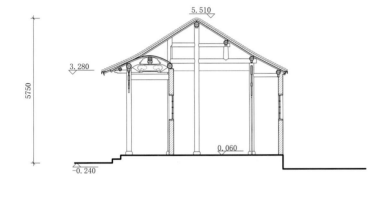

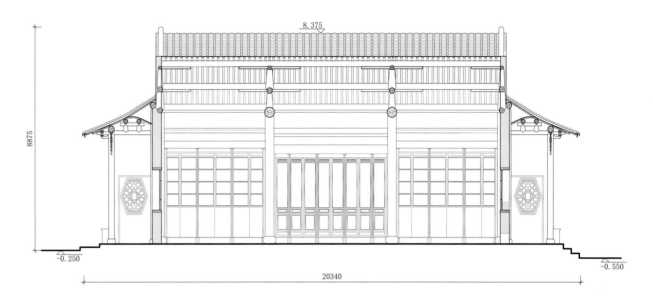

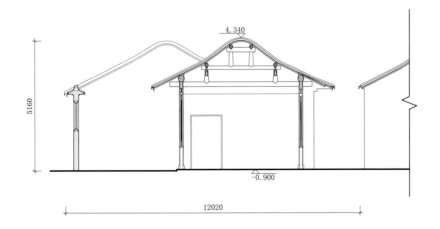

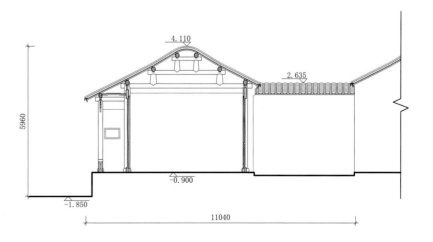

苏州沧浪亭明道堂—瑶华境界—看山楼 C-C 剖面图
Section C-C of Mingdao Hall, Yaohua jingjie, Kanshan Building of Canglangting, Suzhou

苏州沧浪亭明道堂—瑶华境界—看山楼 D-D 剖面图
Section D-D of Mingdao Hall, Yaohua jingjie, Kanshan Building of Canglangting, Suzhou

苏州沧浪亭明道堂—瑶华境界—看山楼 E-E 剖面图
Section E-E of Mingdao Hall, Yaohua jingjie, Kanshan Building of Canglangting, Suzhou

苏州沧浪亭明道堂—瑶华境界—看山楼 F-F 剖面图
Section F-F of Mingdao Hall, Yaohua jingjie, Kanshan Building of Canglangting, Suzhou

0 1 2 3m

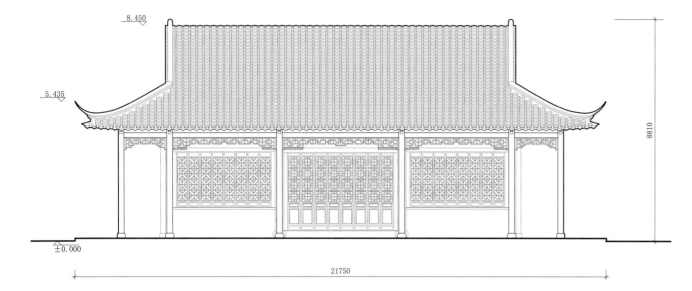

苏州沧浪亭明道堂北立面图
North elevation of Mingdao Hall of Canglangting, Suzhou

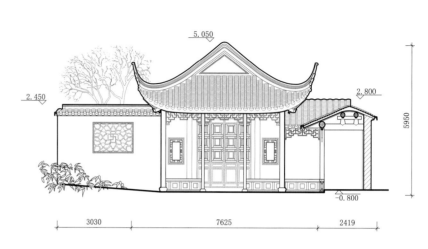

苏州沧浪亭歇山小屋西立面图
West elevation of Xieshan xiaowu of Canglangting, Suzhou

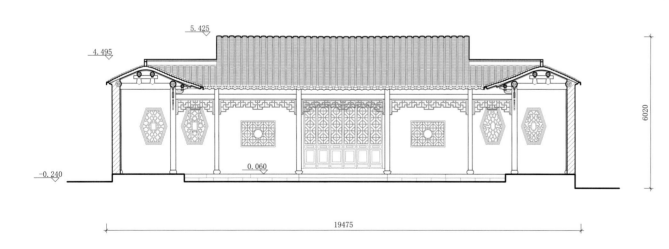

苏州沧浪亭瑶华境界北立面图
North elevation of Yaohua jingjie of Canglangting, Suzhou

苏州沧浪亭闻妙香室西立面图
West elevation of Wenmiao xiangshi of Canglangting, Suzhou

0 1 2 3m

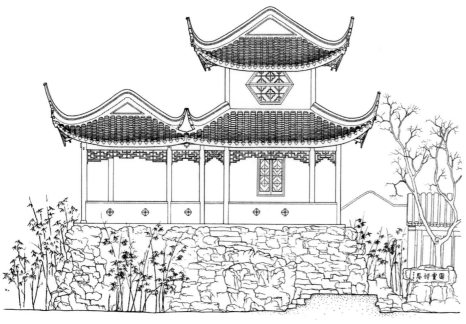

苏州沧浪亭看山楼立面图
Elevation of Kanshan Building of Canglangting, Suzhou

苏州沧浪亭看山楼剖面图
Section of Kanshan Building of Canglangting, Suzhou

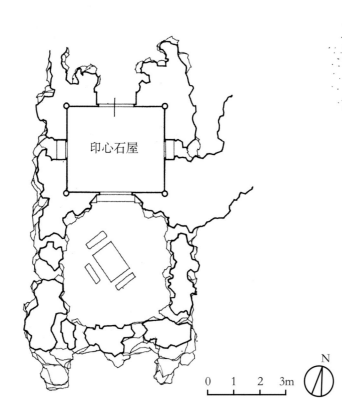

印心石屋

苏州沧浪亭看山楼底层石洞平面图
Plan of ground floor cave of Kanshan Building of Canglangting, Suzhou

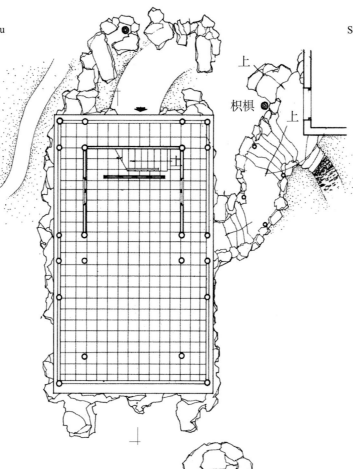

0 1 2 3m N

苏州沧浪亭看山楼一层平面图
Plan of first floor of Kanshan Building of Canglangting, Suzhou

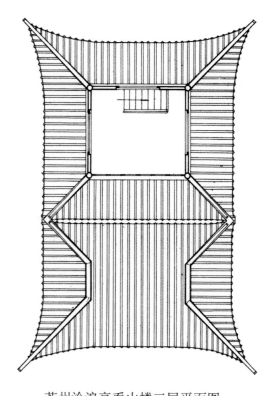

苏州沧浪亭看山楼二层平面图
Plan of second floor of Kanshan Building of Canglangting, Suzhou

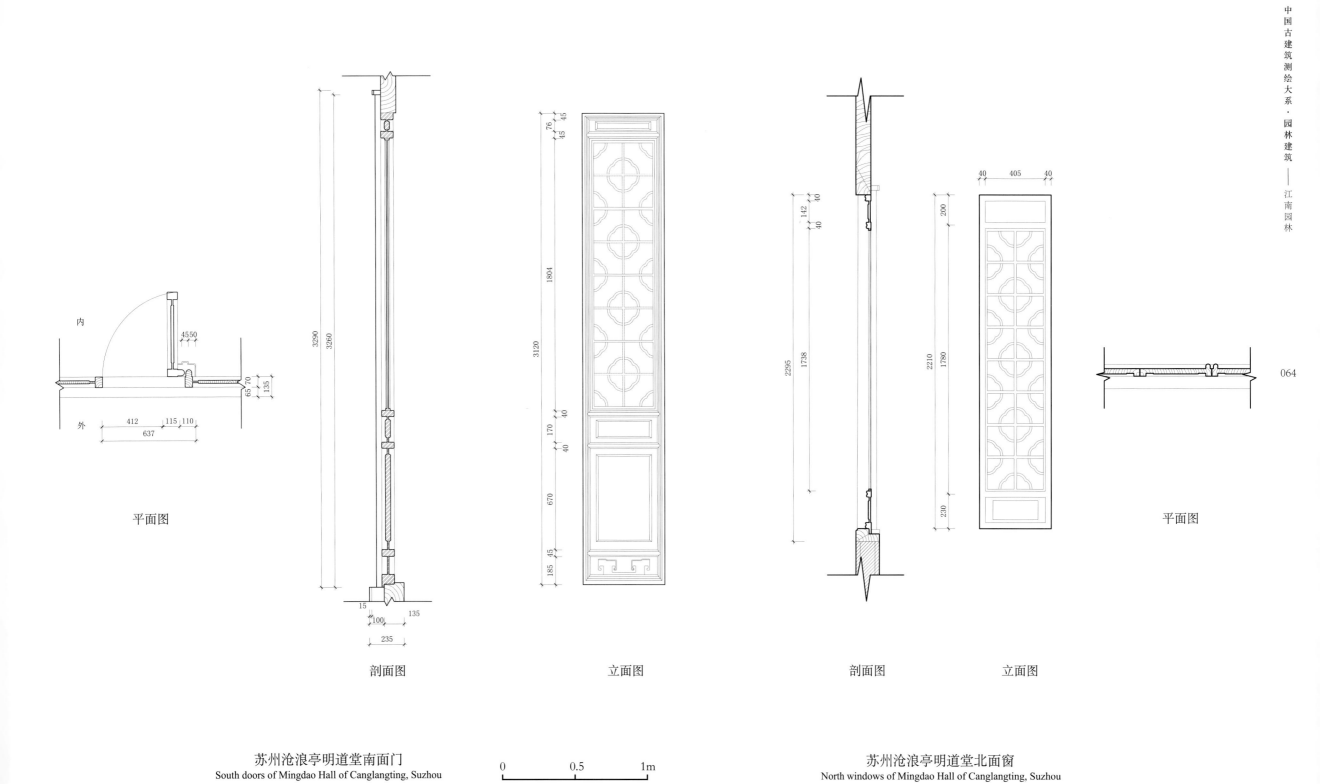

内

外

平面图

剖面图

立面图

剖面图

立面图

平面图

苏州沧浪亭明道堂南面门
South doors of Mingdao Hall of Canglangting, Suzhou

0   0.5   1m

苏州沧浪亭明道堂北面窗
North windows of Mingdao Hall of Canglangting, Suzhou

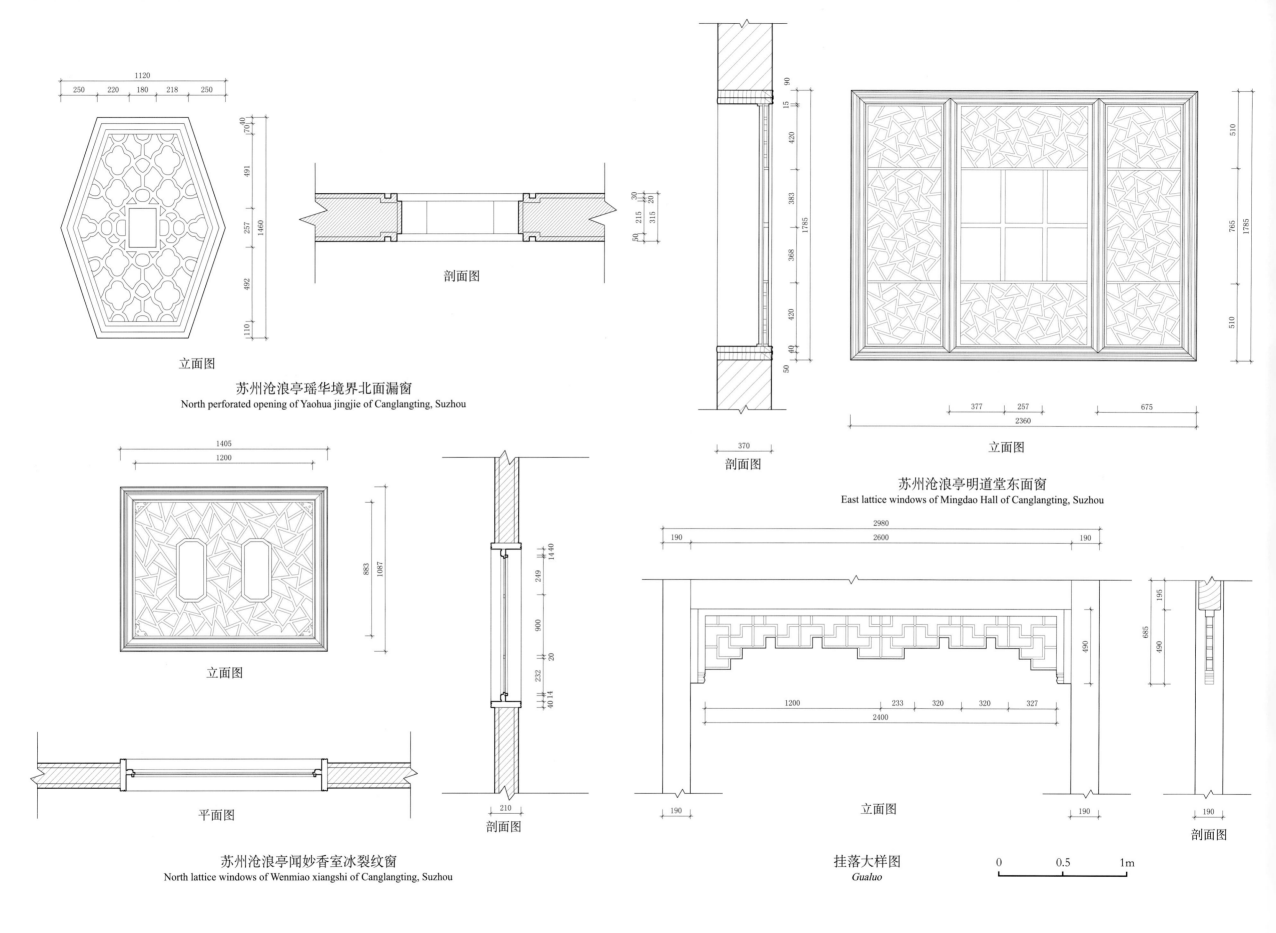

立面图

剖面图

苏州沧浪亭瑶华境界北面漏窗
North perforated opening of Yaohua jingjie of Canglangting, Suzhou

剖面图

立面图

苏州沧浪亭明道堂东面窗
East lattice windows of Mingdao Hall of Canglangting, Suzhou

立面图

平面图

剖面图

苏州沧浪亭闻妙香室冰裂纹窗
North lattice windows of Wenmiao xiangshi of Canglangting, Suzhou

立面图

剖面图

挂落大样图
Gualuo

0   0.5   1m

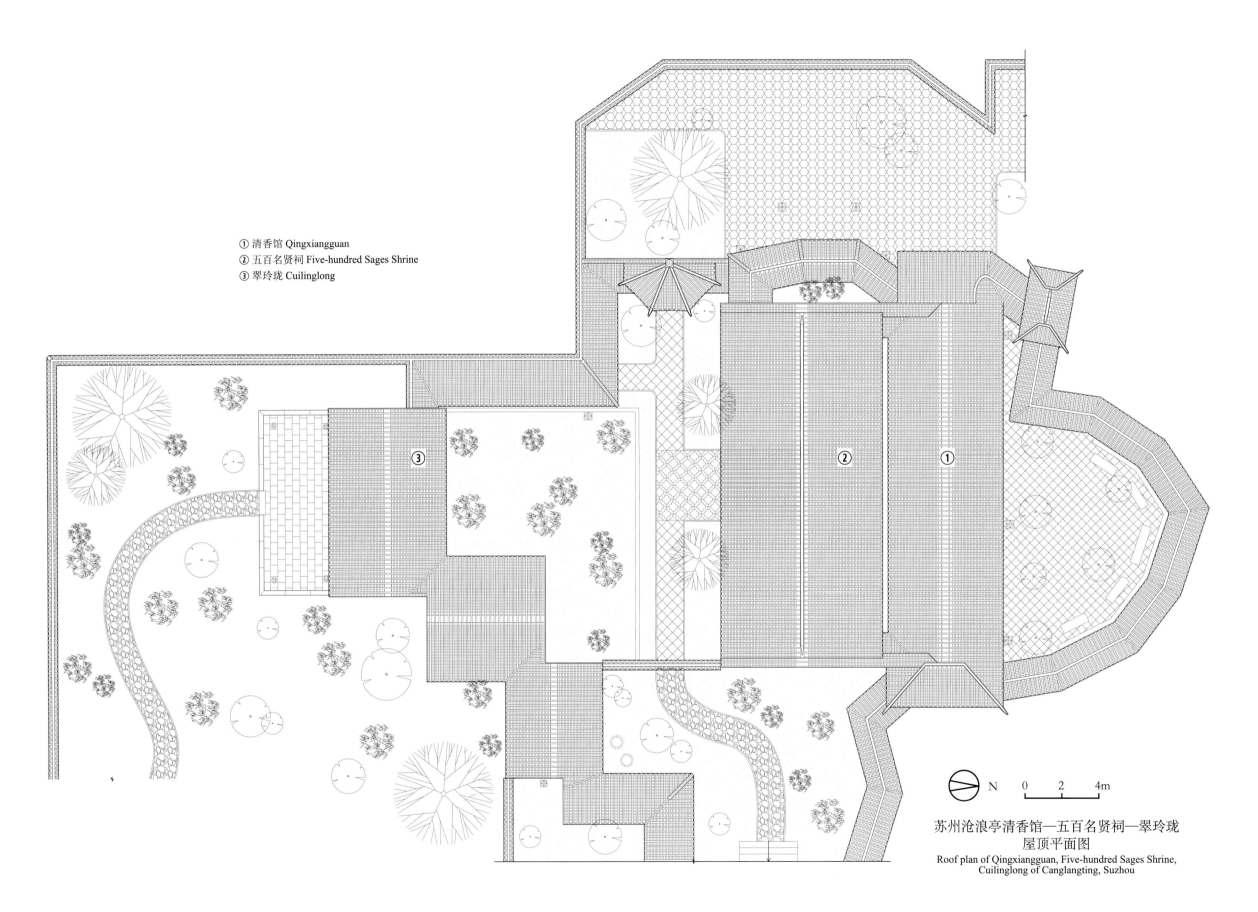

① 清香馆 Qingxiangguan
② 五百名贤祠 Five-hundred Sages Shrine
③ 翠玲珑 Cuilinglong

N   0   2   4m

苏州沧浪亭清香馆—五百名贤祠—翠玲珑
屋顶平面图
Roof plan of Qingxiangguan, Five-hundred Sages Shrine,
Cuilinglong of Canglangting, Suzhou

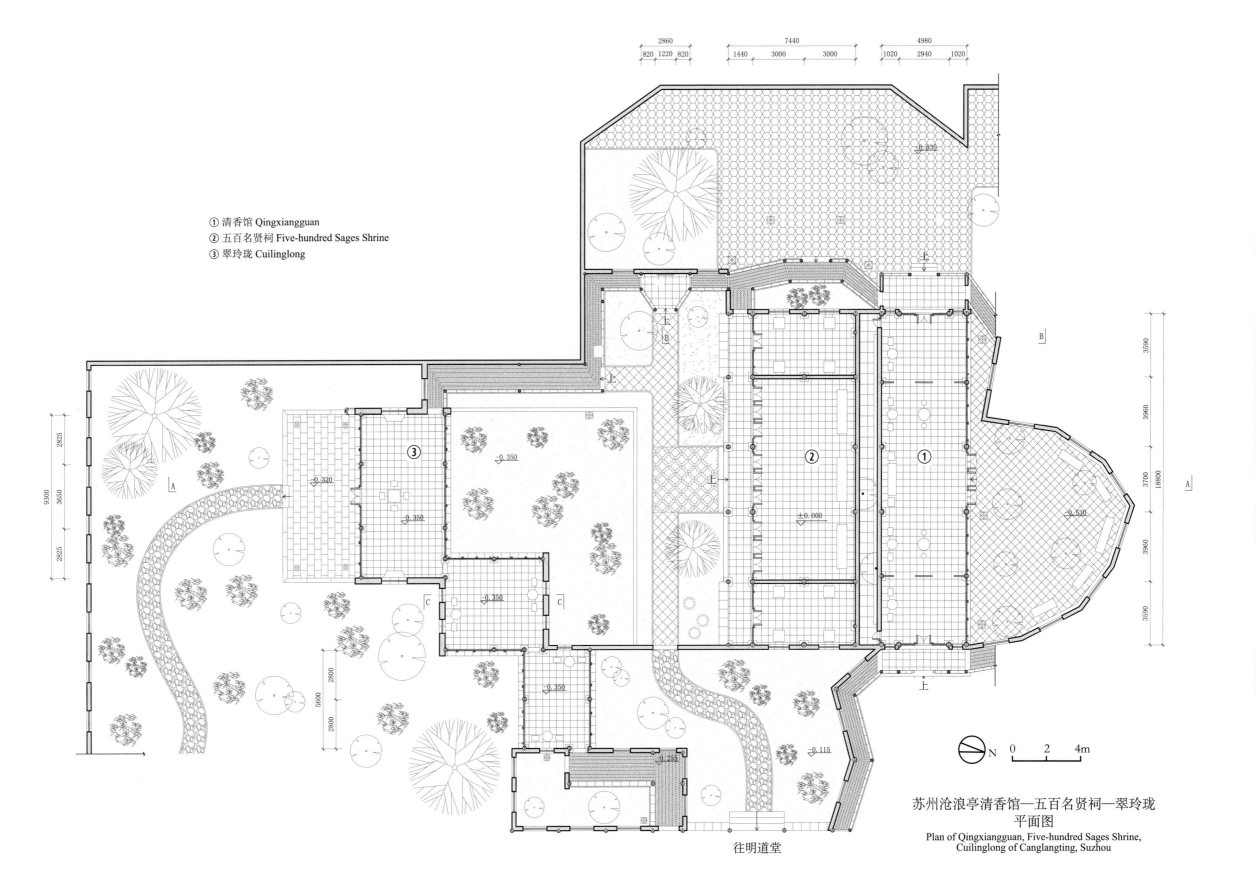

① 清香馆 Qingxiangguan
② 五百名贤祠 Five-hundred Sages Shrine
③ 翠玲珑 Cuilinglong

往明道堂

苏州沧浪亭清香馆—五百名贤祠—翠玲珑
平面图
Plan of Qingxiangguan, Five-hundred Sages Shrine,
Cuilinglong of Canglangting, Suzhou

N   0   2   4m

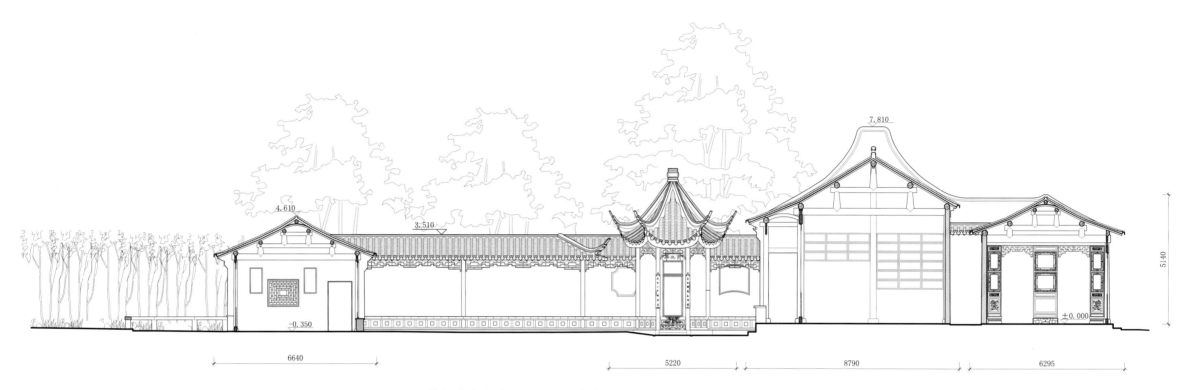

苏州沧浪亭清香馆—五百名贤祠—翠玲珑 A-A 剖面图

Section A-A of Qingxiangguan, Five-hundred Sages Shrine, Cuilinglong of Canglangting, Suzhou

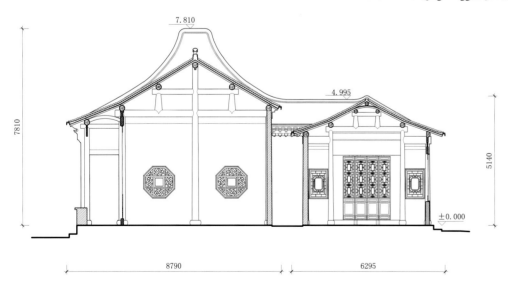

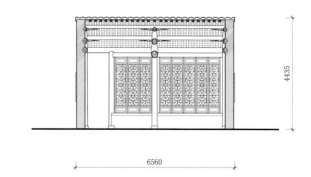

苏州沧浪亭清香馆—五百名贤祠—翠玲珑 B-B 剖面图

Section B-B of Qingxiangguan, Five-hundred Sages Shrine, Cuilinglong of Canglangting, Suzhou

苏州沧浪亭清香馆—五百名贤祠—翠玲珑 C-C 剖面图

Section C-C of Qingxiangguan, Five-hundred Sages Shrine, Cuilinglong of Canglangting, Suzhou

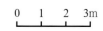

0 1 2 3m

18800
3900　3650　3700　3650　3900

630 1020
630
1680　4980
1020 630

1500
1500　7440
1500
1440

3590　3960　3700　3960　3590
18800

0　1　2　3m

N

苏州沧浪亭清香馆—五百名贤祠仰视平面图
Reflected ceiling plan of Qingxiangguan and Five-hundred Sages Shrine of Canglangting, Suzhou

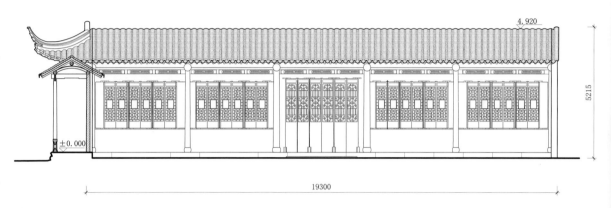

4,920
5215
±0.000
19300

苏州沧浪亭清香馆北立面图
North elevation of Qingxiangguan of Canglangting, Suzhou

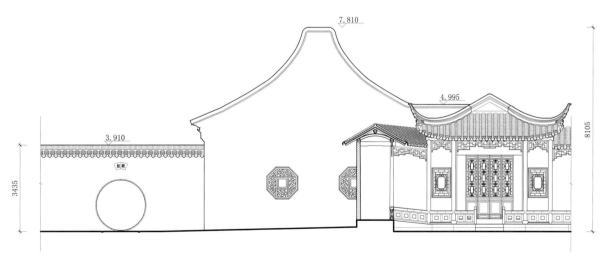

7,810
4,995
3,910
8105
3435

苏州沧浪亭清香馆东立面图
East elevation of Qingxiangguan of Canglangting, Suzhou

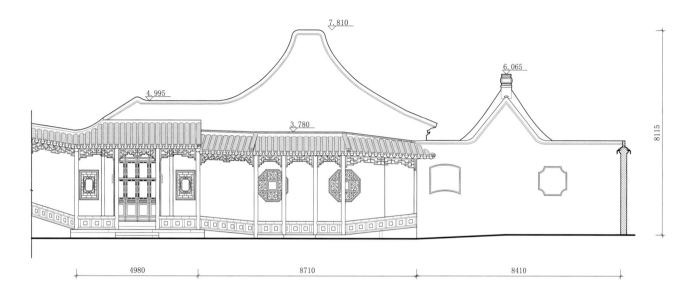

苏州沧浪亭清香馆西立面图
West elevation of Qingxiangguan of Canglangting, Suzhou

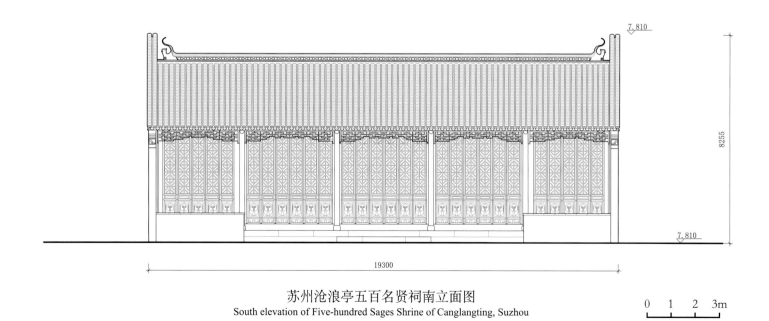

苏州沧浪亭五百名贤祠南立面图
South elevation of Five-hundred Sages Shrine of Canglangting, Suzhou

0 1 2 3m

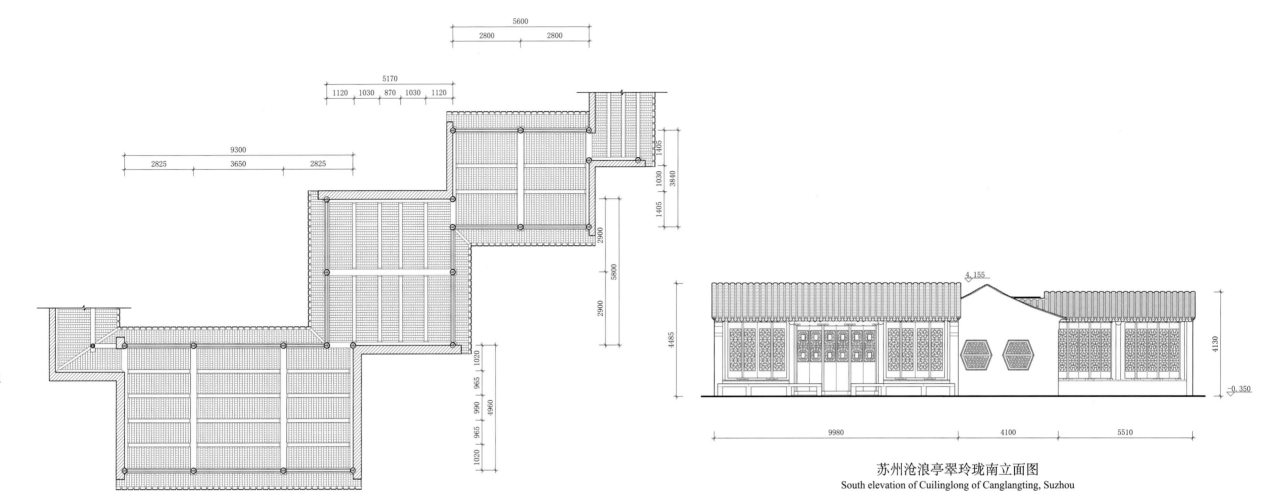

苏州沧浪亭翠玲珑南立面图
South elevation of Cuilinglong of Canglangting, Suzhou

苏州沧浪亭翠玲珑仰视平面图
Reflected ceiling plan of Cuilinglong of Canglangting, Suzhou

N

0 1 2 3m

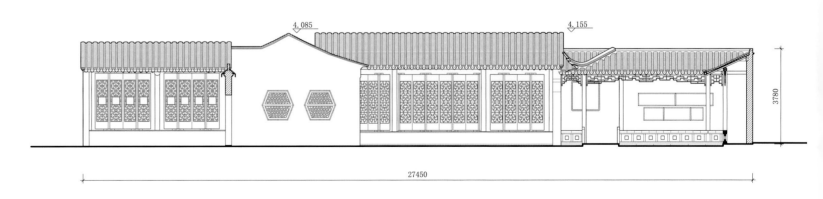

苏州沧浪亭翠玲珑北立面图
North elevation of Cuilinglong of Canglangting, Suzhou

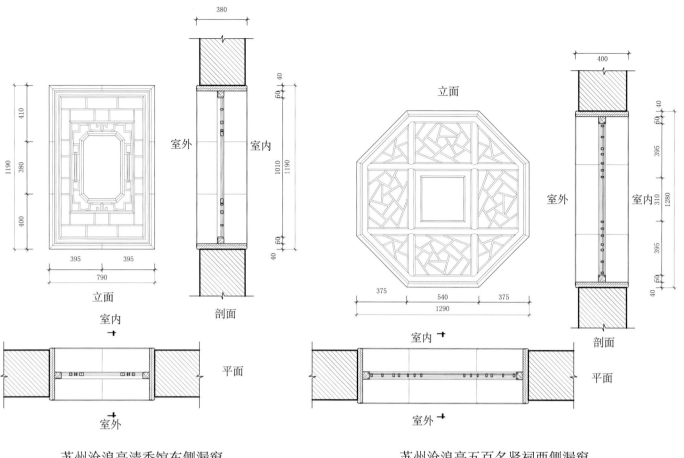

苏州沧浪亭清香馆东侧漏窗
East lattice window of Qingxiangguan of Canglangting, Suzhou

苏州沧浪亭五百名贤祠西侧漏窗
West lattice window of Five-hundred Sages Shrine of Canglangting, Suzhou

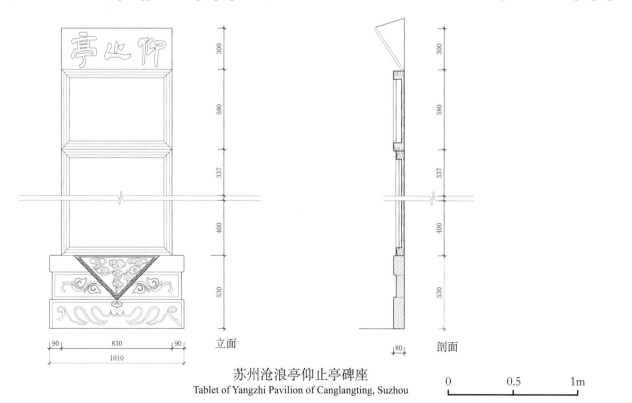

苏州沧浪亭仰止亭碑座
Tablet of Yangzhi Pavilion of Canglangting, Suzhou

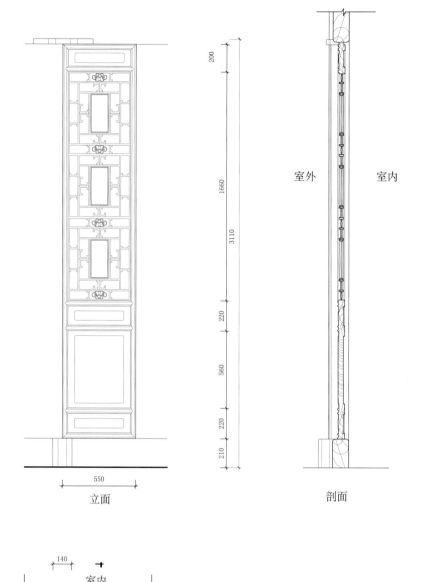

苏州沧浪亭清香馆东门长窗
East window of Qingxiangguan of Canglangting, Suzhou

0　　0.5　　1m

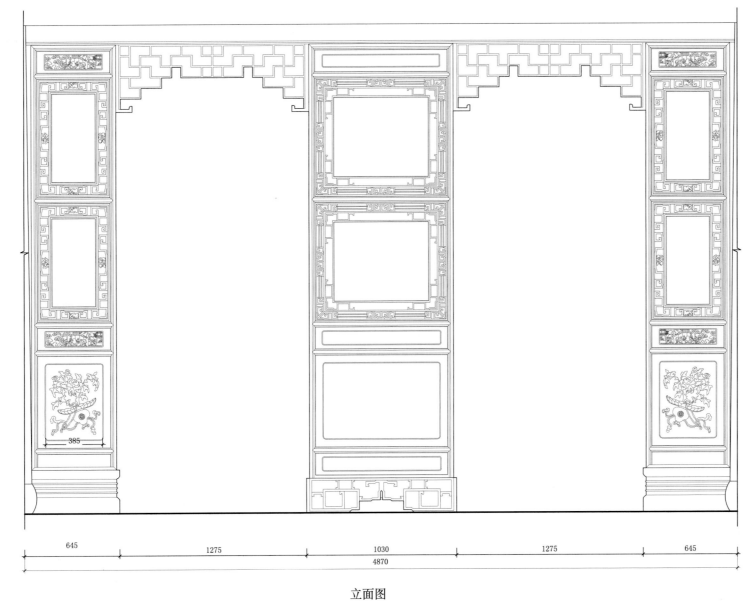

立面图

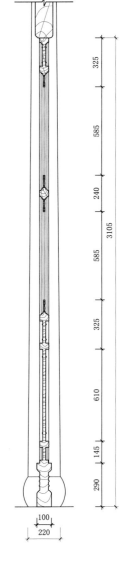

剖面图

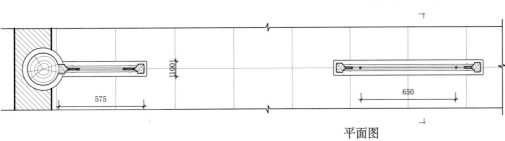

平面图

苏州沧浪亭清香馆屏风
Interior partitions of Qingxiangguan of Canglangting, Suzhou

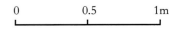

0   0.5   1m

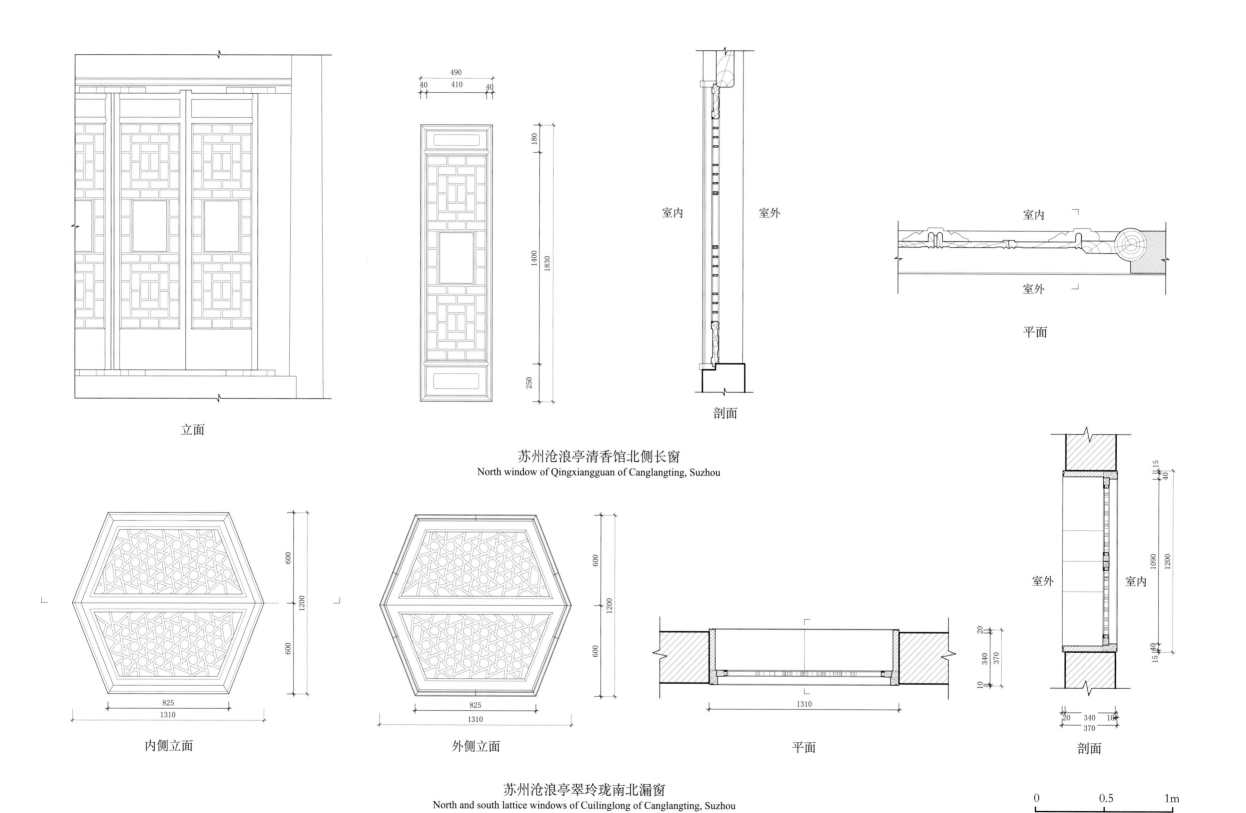

室内　室外

平面

剖面

立面

490
40　410　40

180
1400
1830
250

室内　室外

室内　室外

苏州沧浪亭清香馆北侧长窗
North window of Qingxiangguan of Canglangting, Suzhou

600
1200
600

825
1310

内侧立面

600
1200
600

825
1310

外侧立面

20
340
370
10
1310

平面

15 40
1090
1200
15 40
室外　室内
20　340　10
370

剖面

苏州沧浪亭翠玲珑南北漏窗
North and south lattice windows of Cuilinglong of Canglangting, Suzhou

0　　　0.5　　　1m

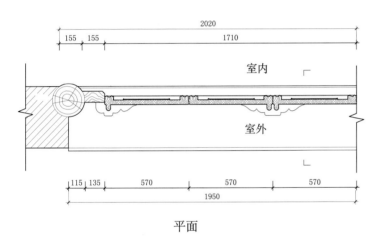

平面

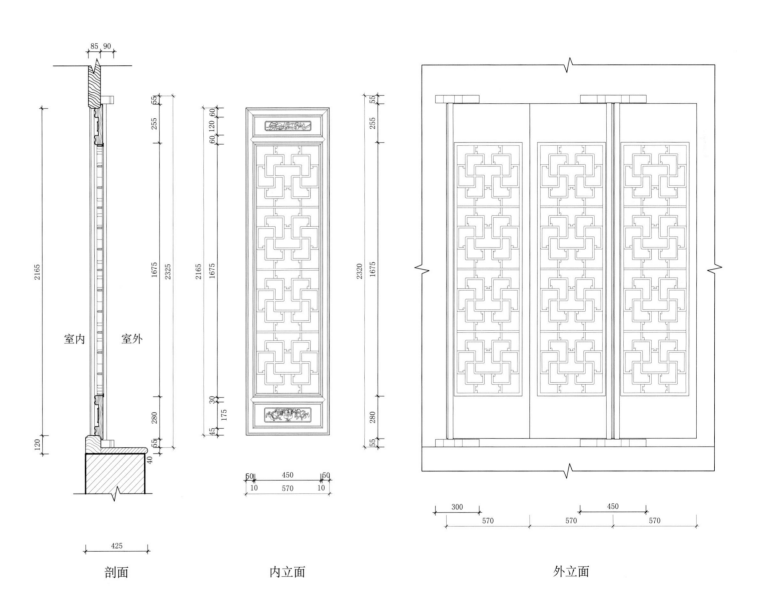

剖面　　　　内立面　　　　　　　外立面

苏州沧浪亭翠玲珑长窗
Windows of Cuilinglong of Canglangting, Suzhou

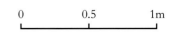

0　　　0.5　　　1m

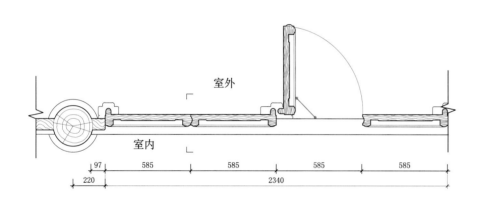

室外

室内

平面

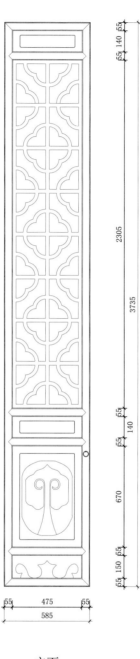

立面

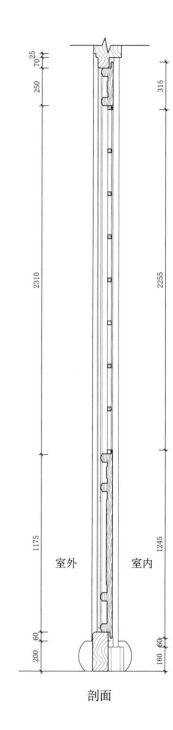

室外　室内

剖面

苏州沧浪亭五百名贤祠长窗
Window-doors of Five-hundred Sages Shrine of Canglangting, Suzhou

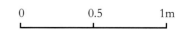

0　　　0.5　　　1m

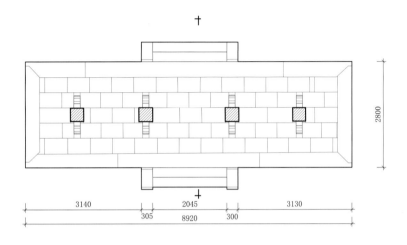

苏州沧浪亭沧浪胜迹坊平面图

Site plan of Canglang shengji archway of Canglangting, Suzhou

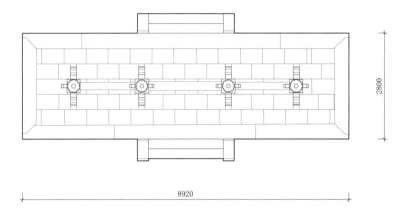

苏州沧浪亭沧浪胜迹坊顶视图

Plan as seen from above of Canglang shengji archway of Canglangting, Suzhou

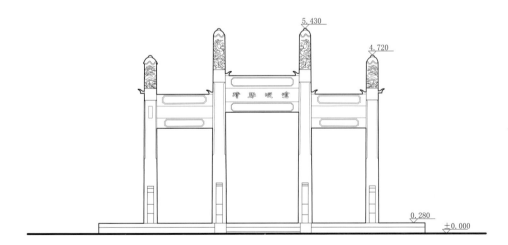

苏州沧浪亭沧浪胜迹坊立面图

Elevation of Canglang shengji archway of Canglangting, Suzhou

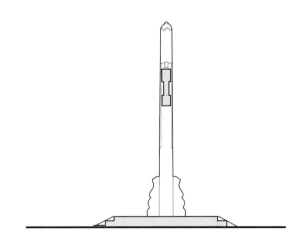

苏州沧浪亭沧浪胜迹坊剖面图

Section of Canglang shengji archway of Canglangting, Suzhou

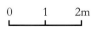

Wangshi Garden, Suzhou

Wangshi Garden (Net Master's Garden) sits at the first lane of Nankuojia in Suzhou. It used to be the site of Wanjuan (Ten-thousand Scrolls) Hall owned by SHI Zhengzhi, a cabinet minister in the Southern Song dynasty. After abandonment, it was rebuilt during the reign of Qing emperor Qianlong and renamed Wangshi. Its subsequent renovations led to the layout people see today.

The garden rests on the western side of its residence, covering half a hectare. The building arrangements divides the garden into two courtyards: the small southern, courtyard contains buildings for living and gathering, comprised of Xiaoshan conggui Pavilion, Daoheguan (Dance Hall), and a music room. The northern courtyard is reserved for literati use, marked by Wufeng (Five Peaks) Library; it is accompanied by Jixu (Meditation) Building, Kansong duhuaxuan (Pine-view Painting Study), and Dianchun Study. At the garden's center sits a square pond over 300 m sq with inlets from southeastern and northwestern corners. Along the pond's edge are hollow niches where the water ripples into, evoking its continuity. The main buildings are aligned along the pond, accompanied by rockeries and flower beds while separated by courtyard and trees. The compact layout amplifies both the verticality and depth of garden scenes without compromising the central scenery.

Wangshi Garden presents a range of notable techniques including the careful use of proportion and contrasts for heightening spatial relationships, as well as concise layout, measured scale, and exquisite scene making. It is the epitome of a classical, medium-sized Suzhou garden.

苏州网师园

网师园在苏州带城桥南阔家头巷，原为南宋史正志『万卷堂』故址，时称『渔隐』，后荒废，清乾隆年间重建，题名『网师』，后几经兴衰，多次修整，成今日布局规模。

此园位于住宅西侧，面积约半公顷。建筑组成庭院两区：南面『小山丛桂轩』与『蹈和馆』、琴室为一区居住、宴聚用的小庭院；北面五峰书屋、集虚斋、看松读画轩、殿春簃等组成以书房为主的庭院一区。中部以三百多平方米的水池为中心，水面略成方形，聚而不分，东南、西北两角伸出水湾，黄石池岸低矮呈洞穴状，使池面有水广波延、源头不尽之意。主要建筑与水池间，亘以假山、花台，隔以庭院、树木，不使逼压池面，亦增加园景的层次与深度。

网师园成功运用比例陪衬关系与对比手法，布局紧凑简洁，尺度斟酌恰当，以精致小巧著称，是苏州中型古典园林的代表。

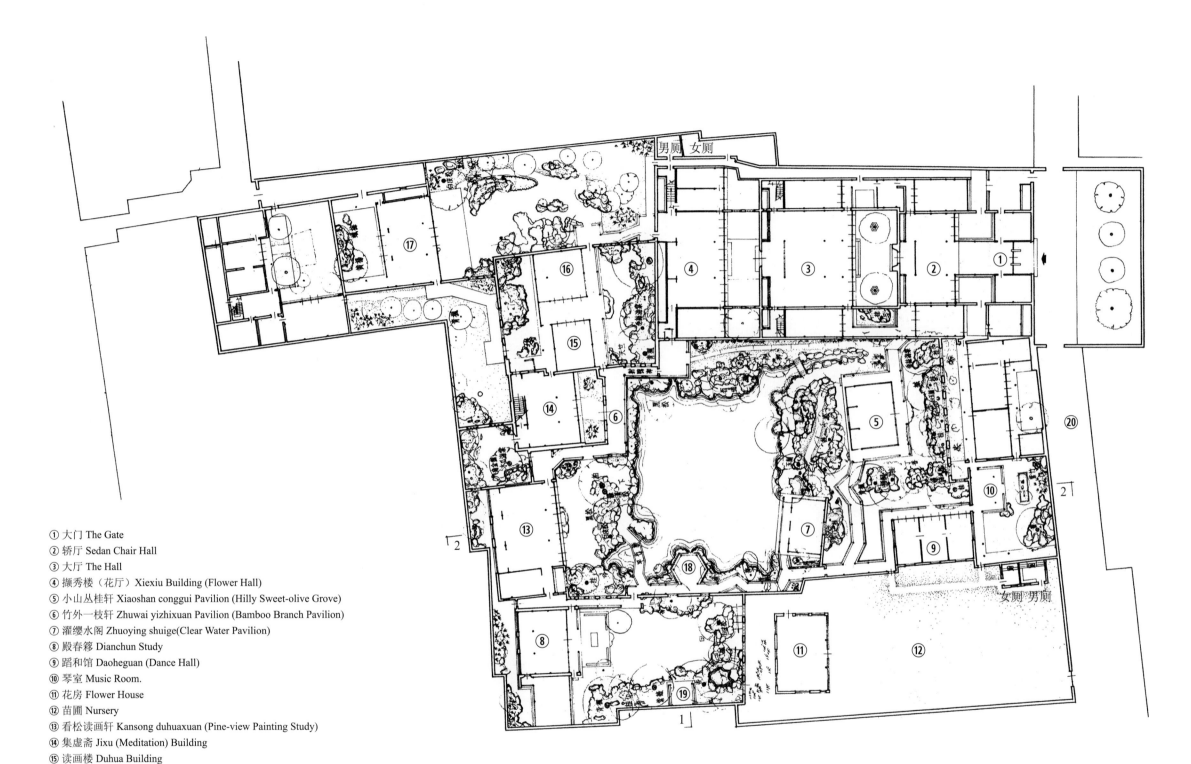

女厕　男厕

① 大门 The Gate

② 轿厅 Sedan Chair Hall

③ 大厅 The Hall

④ 撷秀楼（花厅）Xiexiu Building (Flower Hall)

⑤ 小山丛桂轩 Xiaoshan conggui Pavilion (Hilly Sweet-olive Grove)

⑥ 竹外一枝轩 Zhuwai yizhixuan Pavilion (Bamboo Branch Pavilion)

⑦ 濯缨水阁 Zhuoying shuige(Clear Water Pavilion)

⑧ 殿春簃 Dianchun Study

⑨ 蹈和馆 Daoheguan (Dance Hall)

⑩ 琴室 Music Room.

⑪ 花房 Flower House

⑫ 苗圃 Nursery

⑬ 看松读画轩 Kansong duhuaxuan (Pine-view Painting Study)

⑭ 集虚斋 Jixu (Meditation) Building

⑮ 读画楼 Duhua Building

⑯ 五峰书屋 Wufeng (Five Peaks) Library

⑰ 梯云室 Tiyun Room

⑱ 月到风来亭 Yuedao fenglai Pavilion

⑲ 冷泉亭 Lengquan Pavilion

⑳ 阔家头巷 The first lane of Kuojia

苏州网师园平面图
Site plan of Wangshi Garden, Suzhou

 N

 0　　5　　10m

苏州网师园透视图
Aerial perspective of Wangshi Garden, Suzhou

冷泉亭
Lengquan Pavilion

殿春簃
Dianchun Study

月到风来亭
Yuedao fenglai Pavilion

看松读画轩
Kansong duhuaxuan

集虚斋
Ji xu Building

竹外一枝轩
Zhuwai yizhixuan Pavilion

苏州网师园 1-1 剖面图
Section 1-1 of Wangshi Garden, Suzhou

月到风来亭
Yuedao fenglai Pavilion

看松读画轩
Kansong duhuaxuan

0　　2　　4m

琴室
Music Room

小山丛桂轩
Xiaoshan conggui Pavilion

濯缨水阁
Zhuoying shuige

濯缨水阁
Zhuoying shuige

苏州网师园 2-2 剖面图
Section 2-2 of Wangshi Garden, Suzhou

小山丛桂轩前院
Forecourt of Xiaoshan conggui Pavilion

Xiaoshan conggui (Hilly Sweet-olive Grove) Pavilion sits near Net Master's Garden's southern entrance.

The pavilion resembles a four-sided hall with a panoramic view. The building has Yungang—a yellow rockery hill—to its rear, a wall to its east, and Daoheguan to its west. Facing the pavilion are flower beds made of lake rockeries which creates a mirrored scene of the rear with sweet olive blossom as the focus. Other plants include *yulan*, crape myrtle, maple and wintersweet, which brings changing sceneries by season. A small yard to the west also has flower beds dotted by sweet olive trees, maple, nandina, and Chinese parasol trees; amidst it all, a winding gallery connects the pavilion, Daoheguan and a waterside pavilion beyond. Though limited in space, the courtyard is filled with scenes which unfolds as one strolls from east to west.

小山丛桂轩在网师园南部靠近入口处。

轩略似四面厅，可四面观景。背后有黄石假山「云冈」，东倚住宅界墙，西有蹈和馆。轩前对景为湖石花台，台上植桂花作为主要观赏对象，间植玉兰、紫薇、槭树、蜡梅等花木，四季景色多变。

轩西小院也筑花台，配以桂、槭、天竹、梧桐等花木，中有曲廊逶迤而通蹈和馆及灌缨水阁。此院空间虽小，由东而西层次较多。

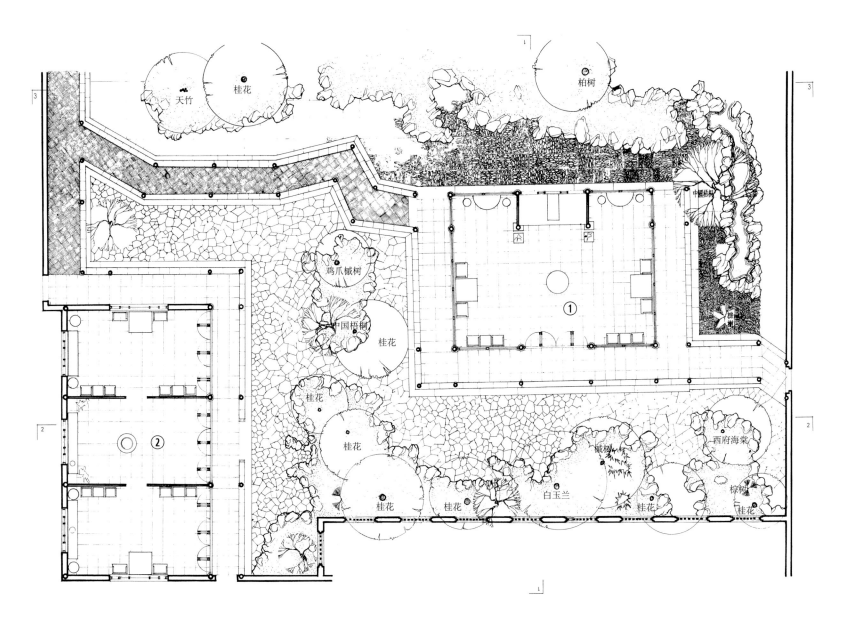

① 小山丛桂轩
Xiaoshan conggui Pavilion (Hilly Sweet-olive Grove)
② 蹈和馆
Daoheguan (Dance Hall)

N 0 1 2 3m

小山丛桂轩前院平面图
Plan of Forcourt of Xiaoshan conggui Pavilion

小山丛桂轩前院 1-1 剖面图
Section 1-1 of Forecourt of Xiaoshanconggui Pavilion

小山丛桂轩前院 2-2 剖面图
Section 2-2 of Forecourt of Xiaoshanconggui Pavilion

小山丛桂轩前院 3-3 剖面图
Section 3-3 of Forecourt of Xiaoshanconggui Pavilion

0　　　1　　　2m

竹外一枝轩
Zhuwai yizhixuan Pavilion

Tucked in Net Master's Garden's northeastern corner, Zhuwai yizhixuan (Bamboo Branch Pavilion) is a joint structure combining a corridor and pavilion. The whitewashed wall and red flowers framed by its openings evoke a solid-void relationship while leaking the courtyard scenes outward. This structure exemplifies the garden's unique circulation space; it is also one of the best spot for viewing the garden's water scenery.

「竹外一枝轩」位于网师园的东北角，是一处廊轩结合的小建筑，造型虚实相间，粉墙与空窗、红花相映成趣，室内空间开敞，是园林内流动空间的典型之作，也是观赏周围景色的最佳点之一。

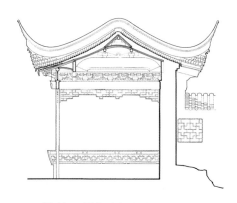

竹外一枝轩剖面图
Section of Zhuwai yizhixuan Pavilion

0 0.5 1m

① 竹外一枝轩
  Zhuwai yizhixuan Pavilion (Bamboo Branch Pavilion)
② 集虚斋
  Jixu (Meditation) Building

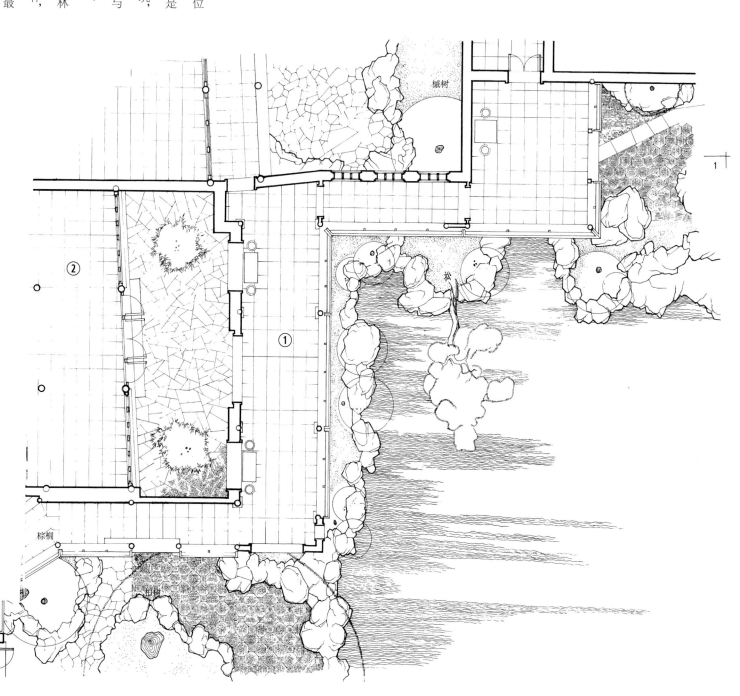

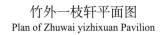

N 0 3 6m

竹外一枝轩平面图
Plan of Zhuwai yizhixuan Pavilion

088

竹外一枝轩南立面图
South elevation of Zhuwai yizhixuan Pavilion

竹外一枝轩西立面图
West elevation of Zhuwai yizhixuan Pavilion

0　1　2　3m

殿春簃
Dianchun Study

殿春簃是网师园西部庭院中的书斋，造型精美，陈设典雅。

「簃」是小屋之意，殿春是晚春的别称，殿春簃意为晚春观赏花卉的次要场所。庭院内种植芍药，芍药花开晚春，又是牡丹的花相，有诗情画意的隐喻。

该处庭院曾被美国纽约大都会博物馆仿造于馆内，称「明轩」，造型略有改变。

Dianchun (End of Spring) Study sits to Net Master's Garden's west. The name of the study implies an alternative venue for flowers viewing in late spring; its small yard is planted with shaoyao (herbaceous peony), which blossoms exactly then. A replica of the study was erected within the courtyard of MOMA in New York with the name Mingxuan, its style slightly modified.

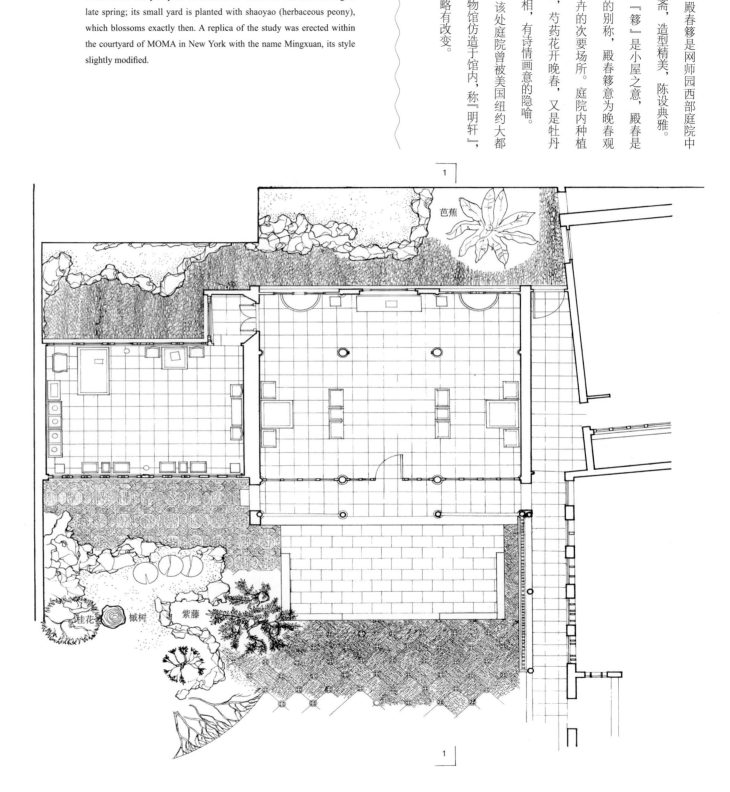

芭蕉

桂花　榔树　紫藤

N

0　1　2　3m

殿春簃平面图
Plan of Dianchun Study

殿春簃南立面图
South elevation of Dianchun Study

0   1   2m

殿春簃剖透视图

Sectional perspective of Dianchun Study

Standing at the southern tip of the water pond, Zhuoying shuige (Clear Water Pavilion) is a single-storied waterside building. It is built with an elegant xieshan (hip-gable) roof with upturned eaves, an outstanding example of sight-seeing architecture. The name of the pavilion suggests a waterside structure. Adjacent to it stands an imposing yellow rockery hill, also a masterpiece of rockery scenes.

濯缨水阁位于网师园水池南端，是一座临水的单层建筑，卷棚歇山顶，嫩戗发戗屋角，造型典雅通透，是观赏类建筑的杰出范例。濯缨水阁意为濯缨之水为清水。水阁旁有雄浑坚实的黄石假山，是假山佳作。

濯缨水阁
Zhuoying shuige

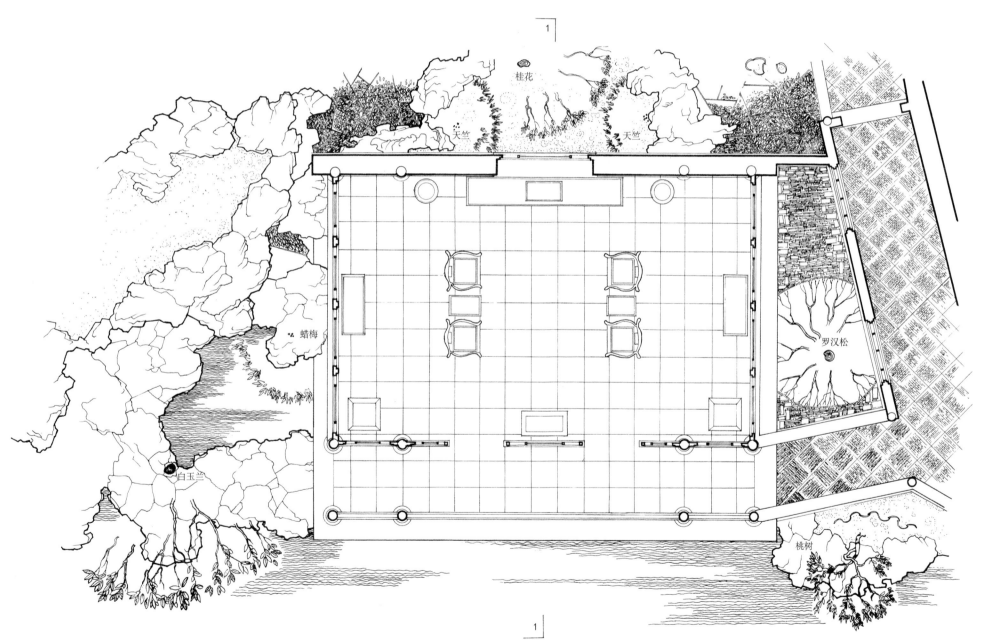

天竺　桂花　天竺

蜡梅

白玉兰

罗汉松

桃树

N

0　1　2m

濯缨水阁平面图
Site plan of Zhuoying shuige

濯缨水阁剖透视图
Sectional perspective of Zhuoying shuige

0   0.5   1m

濯缨水阁北立面图
North elevation of Zhuoying shuige

濯缨水阁东立面图
East elevation of Zhuoying shuige

0   1   2m

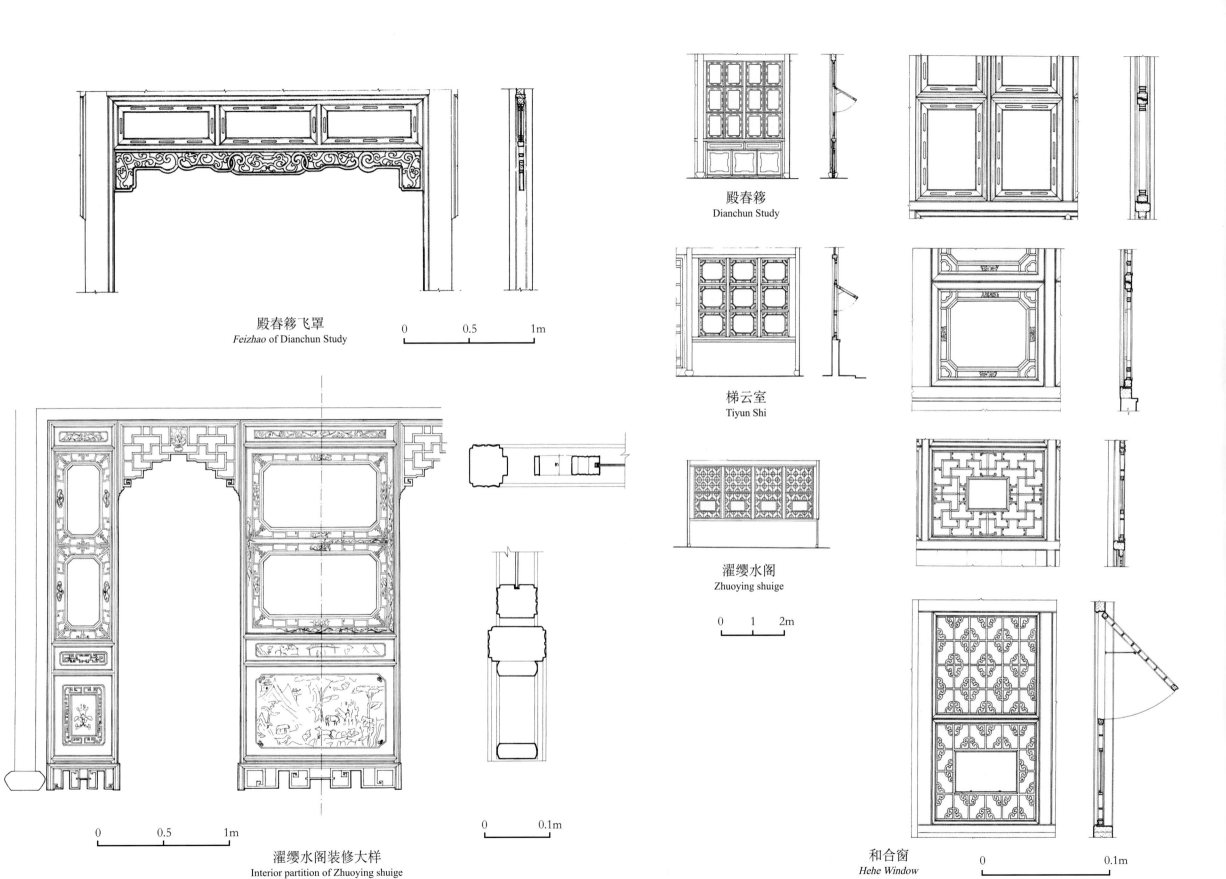

殿春簃飞罩
Feizhao of Dianchun Study

0   0.5   1m

濯缨水阁装修大样
Interior partition of Zhuoying shuige

0   0.5   1m

0   0.1m

殿春簃
Dianchun Study

梯云室
Tiyun Shi

濯缨水阁
Zhuoying shuige

0   1   2m

和合窗
Hehe Window

0   0.1m

Zhuozheng Garden, Suzhou

Sitting at Northeast Street, Loumengnei of Suzhou, Zhuozheng Garden (Humble Administrator's Garden) used to be the private residence of WANG Xianchen, a cabinet minister from the Ming and Qing dynasties (1501-1521). After being neglected for some time, the residence was given a major renovation after the founding of the People's Republic of China. The garden today is divided into three sections spanning a total of 4 ha. Amongst the three sections, the midsection withholds the essence of the garden; it received two rounds of renovations during the reign of emperor Qianlong and Jiaqing—both reusing the original foundations. The garden people see today resembles its appearance from late Qing dynasty: the west section, for instance, inherited its layout during this time. The east section, however, was once a dilapidated site for gardening and only became part of the garden after the founding of the People's Republic of China.

Covering up to 1.2 ha, the garden's midsection is centered on the water. Throughout the garden, the water fragments and clusters, separates yet connects, flowing from the east, west and southwest inlets. To the section's south lies the residential area, where buildings of varying heights are clustered along the water's edge. Here, Yuanxiang (Distant Fragrance) Hall serves as the main activity center; other buildings include Yulantang, Xiaocanglang, Pipayuan, and Haitang chunwu. To the north, the elevated presence of hills, trees, and Jianshan Building completes the garden scenery.

The west section is centered around a meandering pond, which links with the midsection through a hidden niche. The main building, Hall of 36 Mandarin Ducks, sit to the pond's south near the adjacent residential area. To the pond's north spreads a rockery range, upon which stands two pavilions—serving as a corresponding scene for Daoying Building in the northeastern corner.

苏州拙政园

拙政园位于苏州娄门内东北街，创于明正德年间（1501—1521年），为御史王献臣的私园，后屡有兴废。中华人民共和国成立后全面修整与扩建，现园总面积约4公顷，分三部分。中部为全园精华，清乾嘉时期在原有基础上有两次修复，目前建筑基本为清后期面貌；西部现存面貌为清末光绪『补园』时形成；东部『归田园居』旧址，久已荒废，中华人民共和国成立后并入拙政园。

中部约1.2公顷，布局以水为中心。水面有聚有分，分隔变化又彼此贯通，东、西、西南留有水口，伸出如水湾，深远不尽。其南部近住宅，形体不同高低错落的建筑集中于此临水而建，以远香堂为活动中心，周围分布有玉兰堂（花厅）、『小沧浪』（水阁）、枇杷园、海棠春坞等。其北部重山池树木，有见山楼。

西部布局以曲尺形水池为中心，水池与中部有水洞互通，主体建筑『三十六鸳鸯馆』在池南近住宅一侧，池北为假山，山上建宜两亭，与东北隅的倒影楼互为对景。

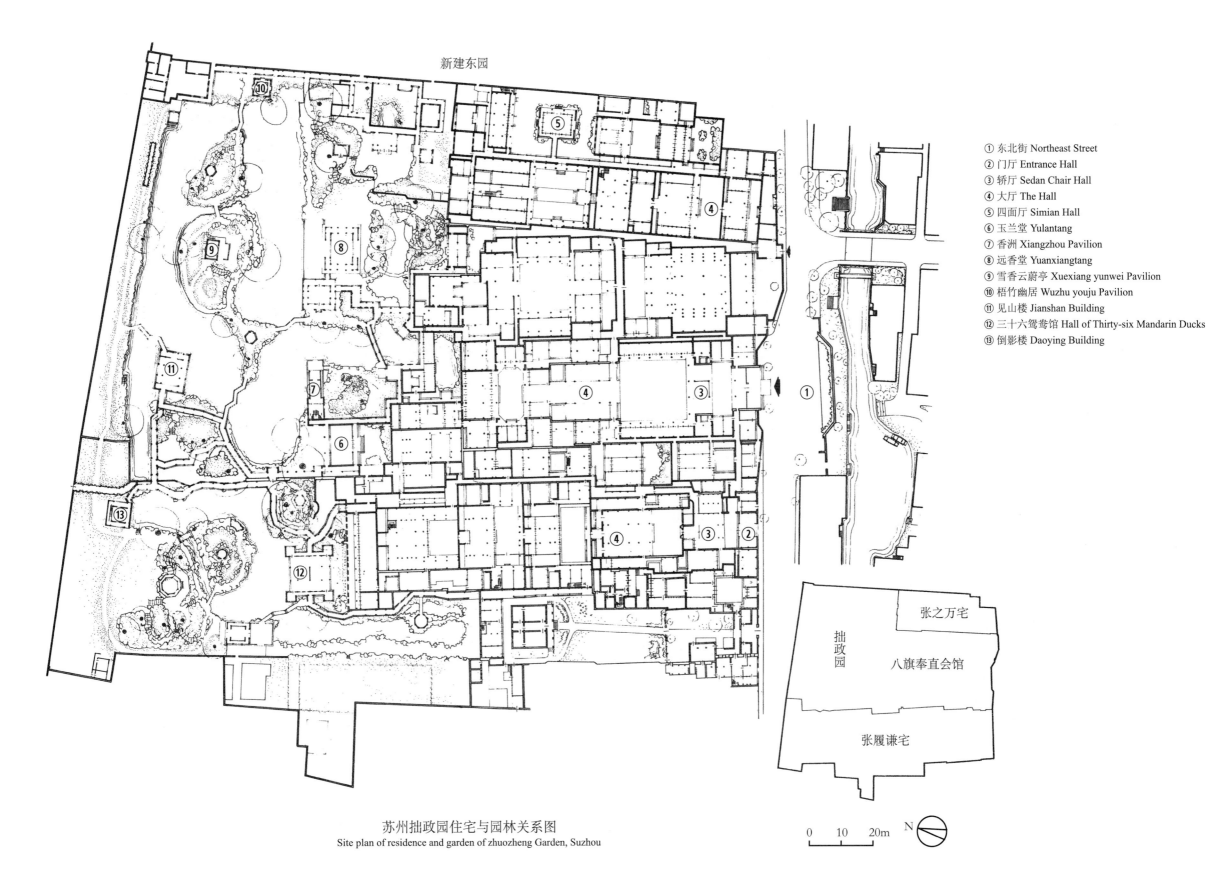

新建东园

① 东北街 Northeast Street
② 门厅 Entrance Hall
③ 轿厅 Sedan Chair Hall
④ 大厅 The Hall
⑤ 四面厅 Simian Hall
⑥ 玉兰堂 Yulantang
⑦ 香洲 Xiangzhou Pavilion
⑧ 远香堂 Yuanxiangtang
⑨ 雪香云蔚亭 Xuexiang yunwei Pavilion
⑩ 梧竹幽居 Wuzhu youju Pavilion
⑪ 见山楼 Jianshan Building
⑫ 三十六鸳鸯馆 Hall of Thirty-six Mandarin Ducks
⑬ 倒影楼 Daoying Building

张之万宅

拙政园

八旗奉直会馆

张履谦宅

苏州拙政园住宅与园林关系图
Site plan of residence and garden of zhuozheng Garden, Suzhou

0 10 20m

N

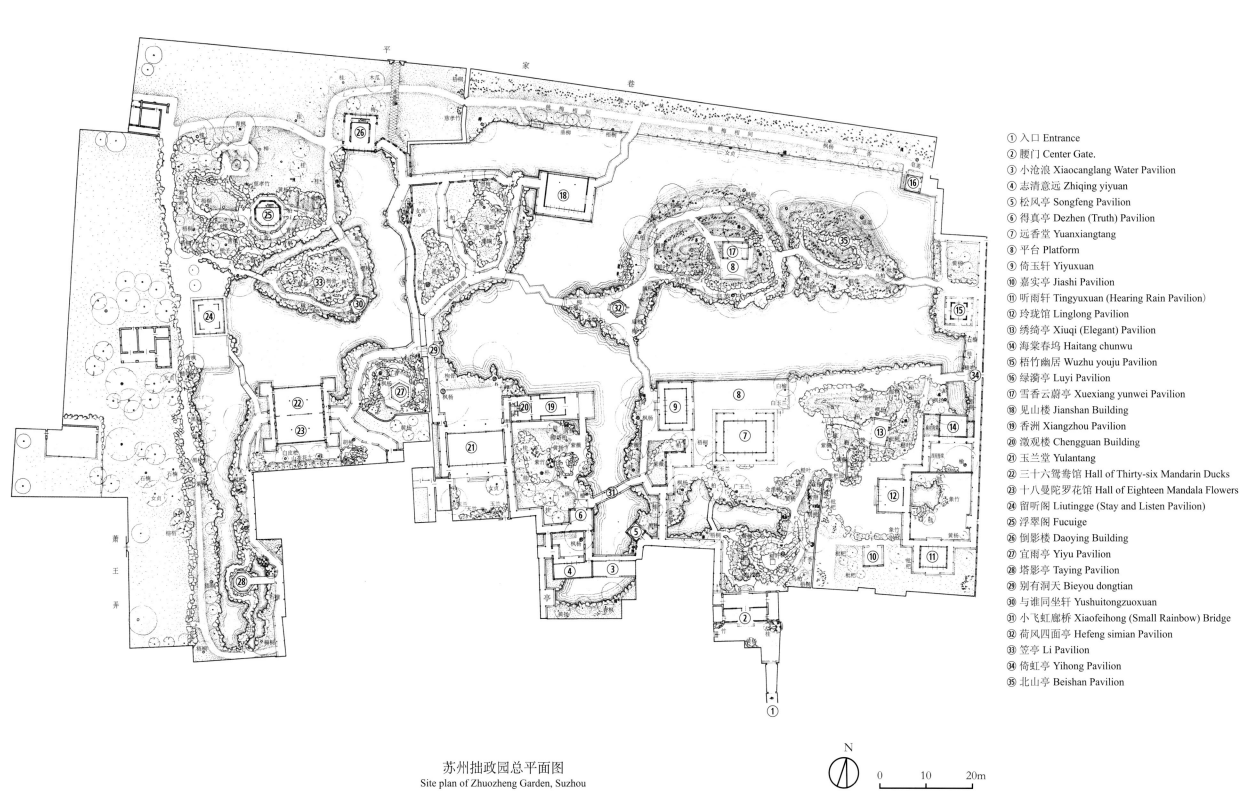

① 入口 Entrance
② 腰门 Center Gate.
③ 小沧浪 Xiaocanglang Water Pavilion
④ 志清意远 Zhiqing yiyuan
⑤ 松风亭 Songfeng Pavilion
⑥ 得真亭 Dezhen (Truth) Pavilion
⑦ 远香堂 Yuanxiangtang
⑧ 平台 Platform
⑨ 倚玉轩 Yiyuxuan
⑩ 嘉实亭 Jiashi Pavilion
⑪ 听雨轩 Tingyuxuan (Hearing Rain Pavilion)
⑫ 玲珑馆 Linglong Pavilion
⑬ 绣绮亭 Xiuqi (Elegant) Pavilion
⑭ 海棠春坞 Haitang chunwu
⑮ 梧竹幽居 Wuzhu youju Pavilion
⑯ 绿漪亭 Luyi Pavilion
⑰ 雪香云蔚亭 Xuexiang yunwei Pavilion
⑱ 见山楼 Jianshan Building
⑲ 香洲 Xiangzhou Pavilion
⑳ 澂观楼 Chengguan Building
㉑ 玉兰堂 Yulantang
㉒ 三十六鸳鸯馆 Hall of Thirty-six Mandarin Ducks
㉓ 十八曼陀罗花馆 Hall of Eighteen Mandala Flowers
㉔ 留听阁 Liutingge (Stay and Listen Pavilion)
㉕ 浮翠阁 Fucuige
㉖ 倒影楼 Daoying Building
㉗ 宜雨亭 Yiyu Pavilion
㉘ 塔影亭 Taying Pavilion
㉙ 别有洞天 Bieyou dongtian
㉚ 与谁同坐轩 Yushuitongzuoxuan
㉛ 小飞虹廊桥 Xiaofeihong (Small Rainbow) Bridge
㉜ 荷风四面亭 Hefeng simian Pavilion
㉝ 笠亭 Li Pavilion
㉞ 倚虹亭 Yihong Pavilion
㉟ 北山亭 Beishan Pavilion

100

苏州拙政园总平面图
Site plan of Zhuozheng Garden, Suzhou

N

0   10   20m

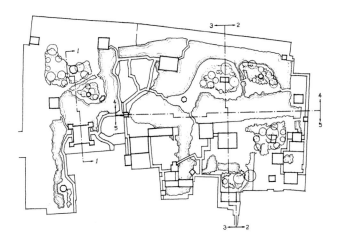

浮翠阁
Fucuige

笠亭
Li Pavilion

与谁同坐轩
Yushuitongzuoxuan

三十六鸳鸯馆
Hall of Thirty-six
Mandarin Ducks

十八曼陀罗花馆
Hall of Eigteen
Mandala Flowers

苏州拙政园 1—1 剖面图
Section 1-1 of Zhuozheng Garden, Suzhou

0　　　　　5　　　　　10m

远香堂
Yuanxiangtang

腰门
Center Gate

0　　　　5　　　　10m

雪香云蔚亭
Xuexiang yunwei Pavilion

绿漪亭
Luyi Pavilion

梧竹幽居
Wuzhu youju Pavilion

倚虹亭
Yihong Pavilion

苏州拙政园 2-2 剖面图
Section 2-2 of Zhuozheng Garden, Suzhou

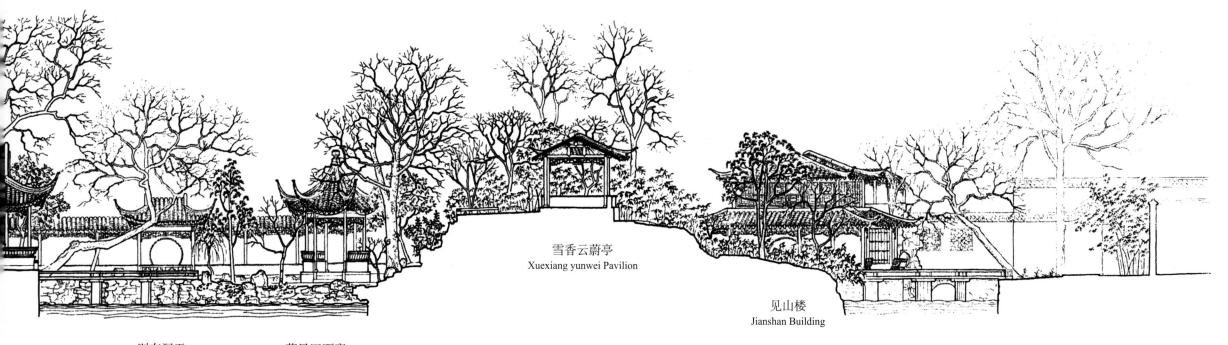

雪香云蔚亭
Xuexiang yunwei Pavilion

见山楼
Jianshan Building

别有洞天
Bieyou dongtian

荷风四面亭
Hefeng simian Pavilion

0   5   10m

腰门
Center Gate

松风亭
Songfeng Pavilion

远香堂
Yuanxiang tang

倚
Yiy

苏州拙政园 3—3 剖面图
Section 3-3 of Zhuozheng Garden, Suzhou

倚玉轩
Yiyuxuan

小飞虹
Xiaofeihong Bridge

香洲
Xiangzhou Pavilion

澄观楼
Chengguan Building

玉兰堂
Yulantang

别有洞天
Bieyou dongtian

0   5   10m

倚虹亭
Yihong Pavilion

海棠春坞
Haitang chunwu

绿漪亭
Luyi Pavilion

枇杷园
Pipa yuan

远香堂
Yuanxiangtang

苏州拙政园 4-4 剖面图
Section 4-4 of Zhuozheng Garden, Suzhou

雪香云蔚亭
Xuexiang yunwei Pavilion

北山亭
Beishan Pavilion

绿漪亭
Luyi Pavilion

梧竹幽居
Wuzhu youju Pavilion

0　　　　　5　　　　　10m

别有洞天
Bieyou dongtian

见山楼
Jianshan Building

荷风四面亭
Hefeng simian Pavilion

苏州拙政园 5-5 剖面图
Section 5-5 of Zhuozheng Garden, Suzhou

Pipa Garden (Loquat Garden) occupies the southeastern corner of Humble Administrator's Garden's midsection, taking on an irregular layout.

Linglong (Delicate) Hall, the main building of Pipayuan, sits east facing west. A gallery leads to Tingyu (Hearing Rain) Pavilion and Haitang chunwu; to the south stands Jiashi Pavilion. Along the north, a rockery hill marks the garden's boundary; topped by Xiuqi (Elegant) Pavilion, lush bamboos and trees, it is the best scene within Loquat Garden. The southwest boundary is defined by an undulating yunqiang (cloud wall) extended from the rockery hill.

Loquat Garden defined space using pavilion, gallery, *yunqiang*, rockery hill trees, and other techniques which leaves the layout open and the scenery stratified. When gazing pass the moon gate of *yunqiang*, one can catch a framed view of Xuexiang yunwei Pavilion from across the water. Conversely, when looking from outside, Jiashi Pavilion becomes the framed object of the corresponding scene.

The courtyard of Loquat Garden is paved with pebbles in an shattered ice pattern to match the latticework of doors and windows in Linglong Pavilion. Consequently, the hall is also dubbed "Jade Ice Pot". The extensive planting of loquat trees led to the naming of Pipa Garden, as well as Jiashi (Fruitful) Pavilion.

枇杷园
Pipa Garden

枇杷园位于拙政园中部东南角，平面不规则。

玲珑馆为主建筑，坐东面西，有廊与「听雨轩」及「海棠春坞」两处庭院相连，南有「嘉实亭」。

北面用假山作屏障，山上筑「绣绮亭」，配以竹丛茂树，构成了庭院中的最佳风景面。西南缭以蜿蜒起伏的云墙，其北端与假山相接。

此院空间分隔采用亭、廊、云墙、假山、树木等手段，平面形式自由，景色层次较多。从云墙上的圆洞门「晚翠」向外看，远处的「雪香云蔚亭」为构图中心形成对景；反之，从圆洞门向内看，又以嘉实亭为主体构成对景。

庭中地面用卵石铺成冰裂纹图案，与玲珑馆门窗上的冰裂纹相映成趣，故玲珑馆又称「玉壶冰」。院内树木主植枇杷，「枇杷园」「嘉实亭」之名由此而来。

① 玲珑馆 Linglong Pavilion
② 绣绮亭 Xiuqi Pavilion
③ 嘉实亭 Jiashi Pavilion

110

枇杷
枇杷
枇杷
柏树
枇杷
枇杷
枇杷
枇杷
枇杷
③
梧桐
枇杷
枇杷
①
枇杷
榉树
柏树
榉树
②
枇杷
枇杷
榉树
榉树
枇杷
芭蕉
石榴

1
2
1
2

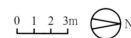

0 1 2 3m　　N

枇杷园平面图
Site plan of Pipa Garden

枇杷园 1-1 剖面图
Section 1-1 of Pipa Garden

枇杷园 2-2 剖面图
Section 2-2 of Pipa Garden

0　1　2m

Yulantang (Yulan Hall) sits in the midsection of Humble Administrator's Garden near the residential area. It used to be a flower hall connected to the residence.

The hall faces a secluded courtyard enclosed by a chamber to the west and walls from east to south. Its impressive scale, lent from its lofty *yingshan* (gable) roof and deep ridge, is comparable to the tall *yulan* trees in front of it. Upon spring's arrival, the trees fully blooms, draping the vicinity in a snowy carpet of white magnolia. Along the southern wall runs a flower bed made of lake rockery, covered by *mudan* (tree peony), bamboos, and rock pinnacles against a serene, whitewashed backdrop.

玉兰堂在拙政园中部靠近住宅一侧，原是一座与住宅相通的花厅。

厅前庭院封闭幽深，西有厢房，东、南两面是围墙。厅堂体形高大、硬山顶，起正脊，厅前庭院内有高大玉兰几株，春天盛放，满树琼瑶，此乃院之胜景。沿南墙筑湖石花台，上植牡丹、竹丛，配以石峰，衬以粉墙，颇为清幽淡雅。

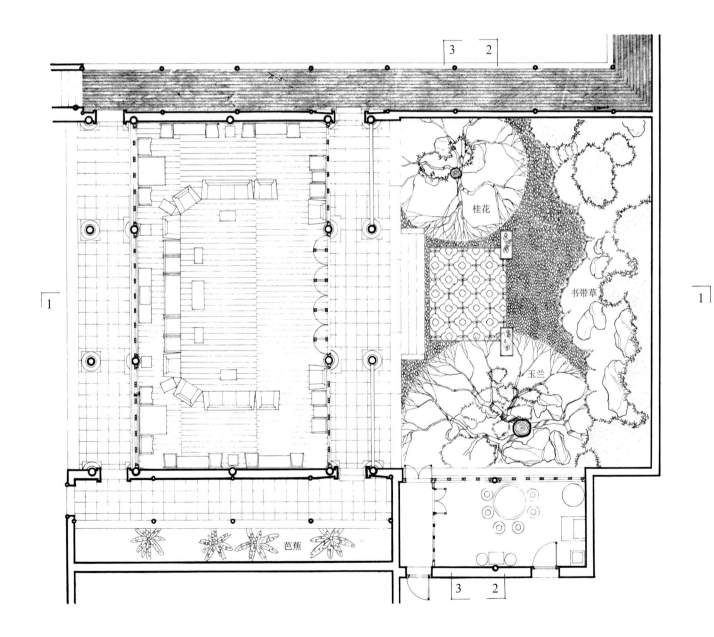

桂花

书带草

玉兰

芭蕉

0 1 2 3m N

玉兰堂庭院平面图
Site plan of Yulantang courtyard

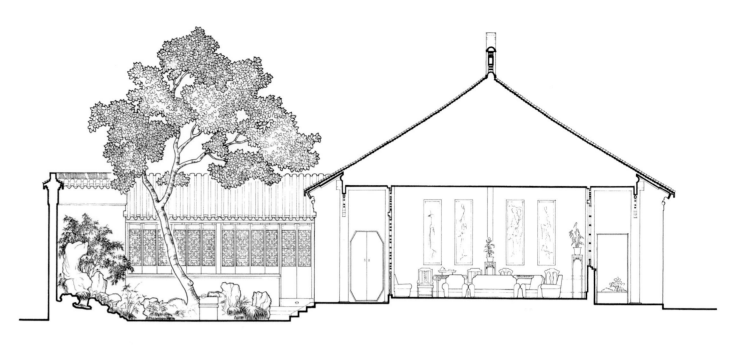

玉兰堂庭院 1-1 剖面图
Section 1-1 of Yulantang courtyard

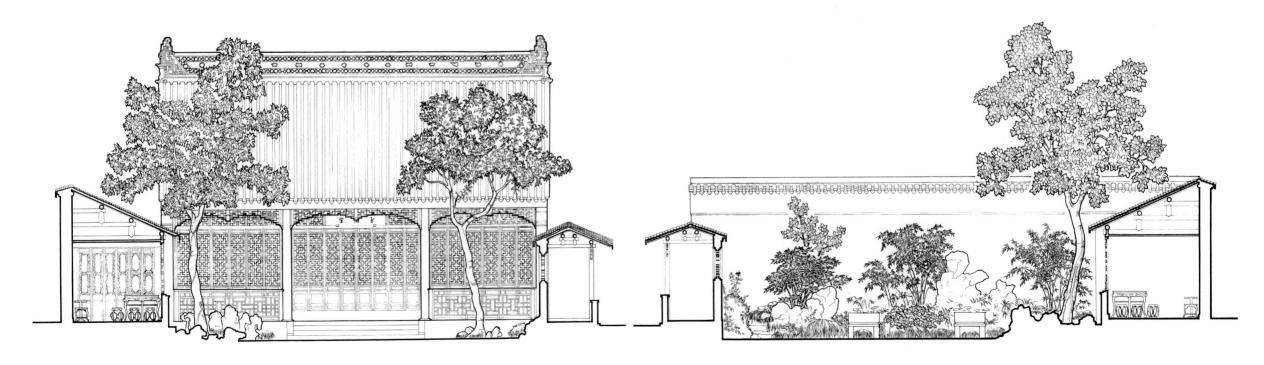

玉兰堂庭院 2-2 剖面图
Section 2-2 of Yulantang courtyard

玉兰堂庭院 3-3 剖面图
Section 3-3 of Yulantang courtyard

0　1　2m

香洲
Xiangzhou Pavilion

Xiangzhou (Fragrant Isle) Pavilion, from Humble Administrator's Garden's midsection, is a waterside structure built to resemble a boat. Its tripartite form is comprised of a tall open front, a lowered midsection, and a two-story back building with an upper floor for distant viewing. When seen from the north, its plain white wall juxtapose against the windows' intricate latticework and the roof's irregular profile—bringing a peculiar yet delightful focus to the garden scene.

香洲位于拙政园中部，是一座临水的仿船形建筑，形似画舫，造型分三段，前舱较高且开敞，中舱低下，尾舱设有二层，可登临远眺。从园林的北部南望，一片粉墙衬托着零星花窗和高低错落的屋顶，构图别其一格。

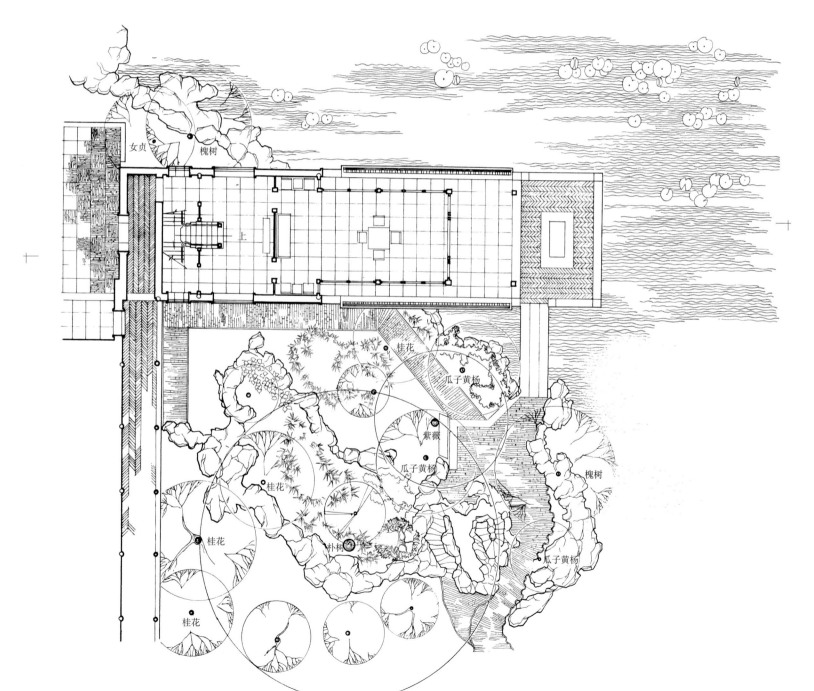

香洲平面图
Site plan of Xiangzhou Pavilion

香洲剖透视图
Sectional perspective of Xiangzhou Pavilion

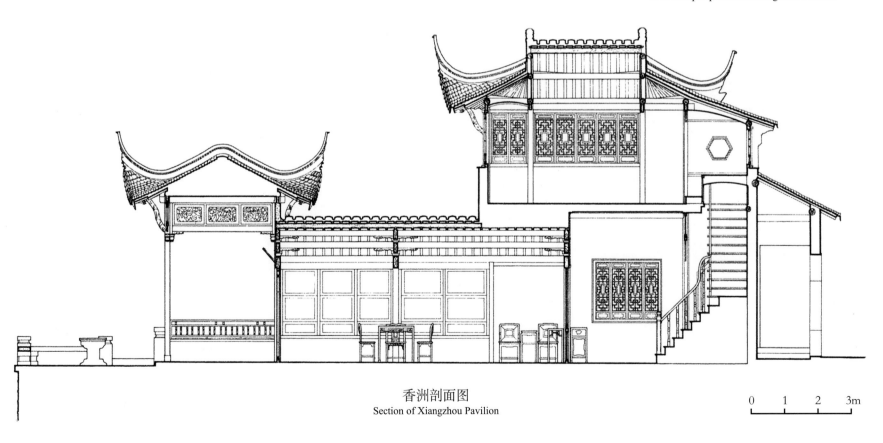

香洲剖面图
Section of Xiangzhou Pavilion

0   1   2   3m

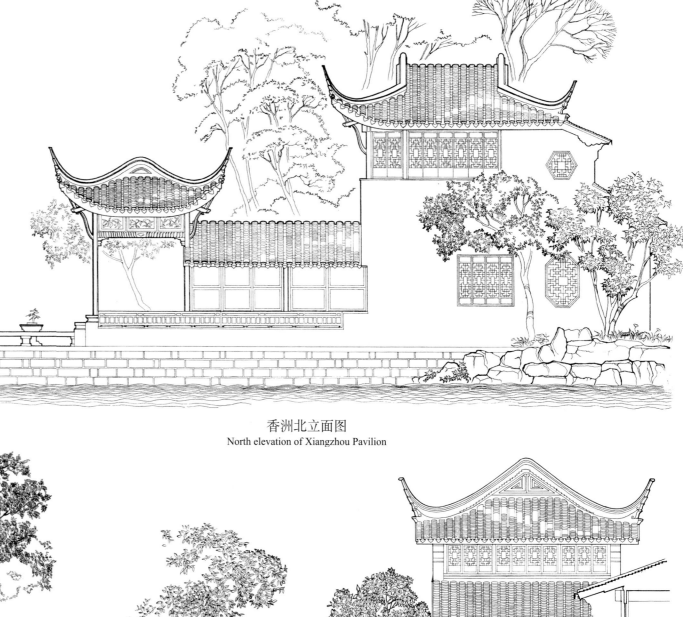

香洲北立面图

North elevation of Xiangzhou Pavilion

香洲东立面图

East elevation of Xiangzhou Pavilion

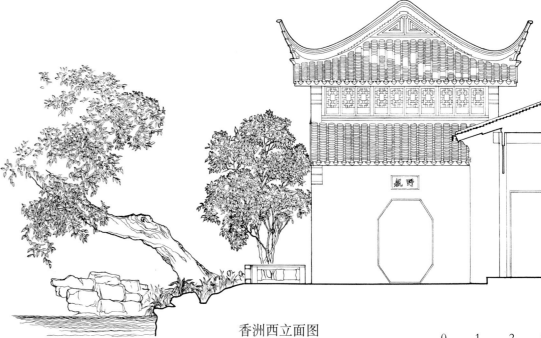

香洲西立面图

West elevation of Xiangzhou Pavilion

0　1　2　3m

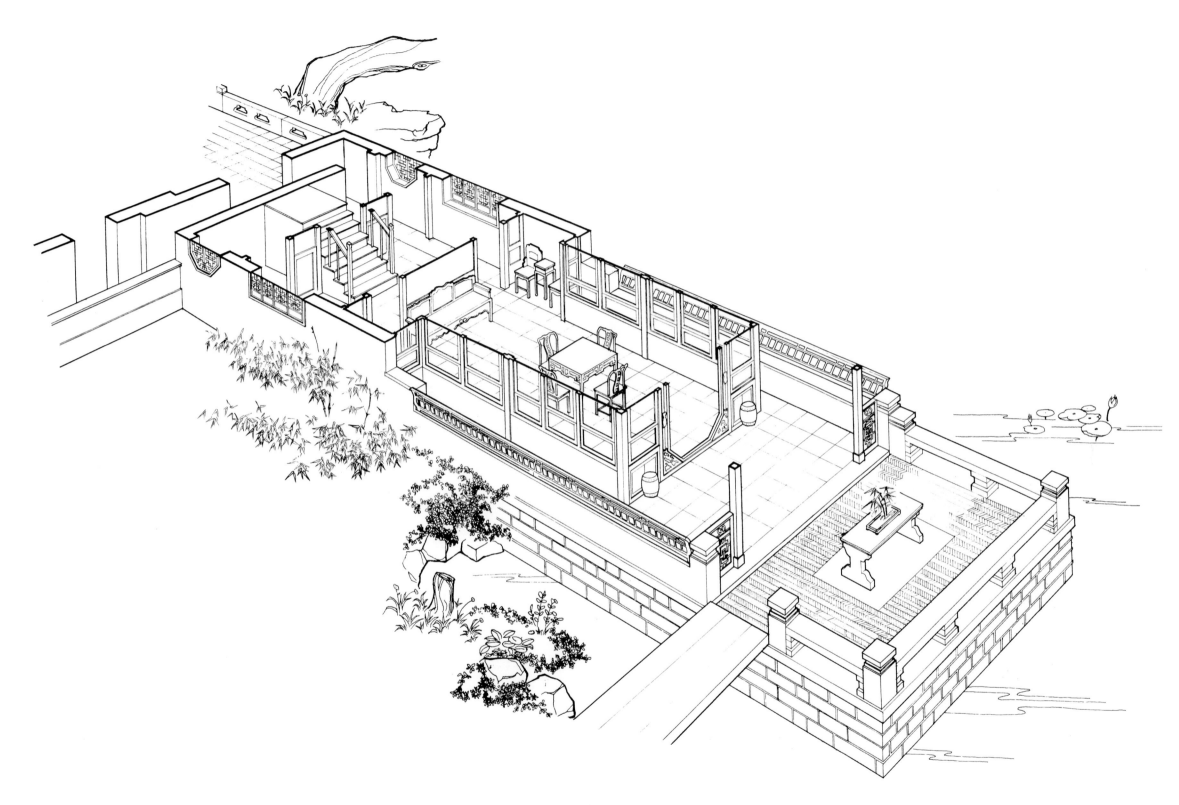

香洲透视图
Perspective of Xiangzhou Pavilion

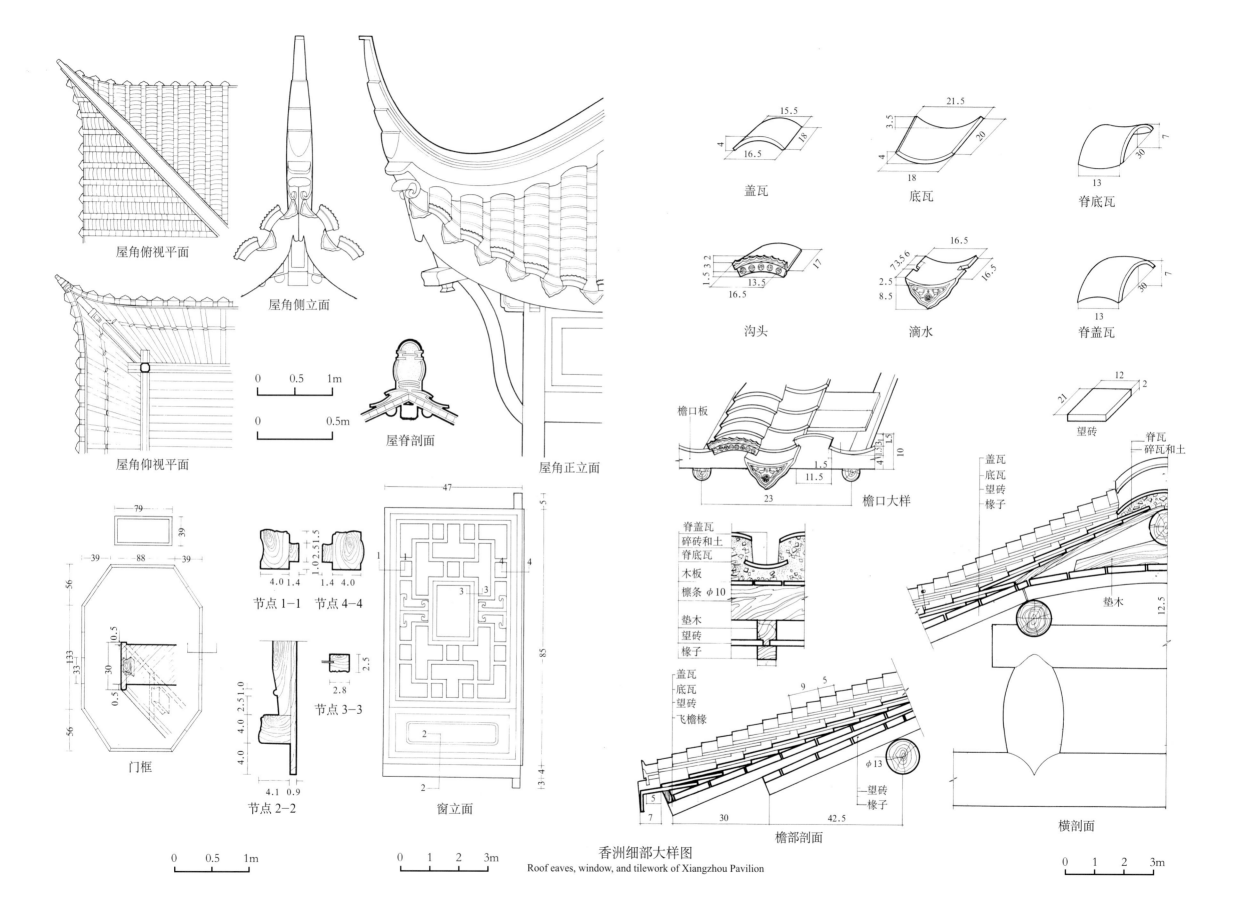

屋角俯视平面

屋角侧立面

屋角仰视平面

屋角正立面

屋脊剖面

门框

节点1—1　节点4—4

节点3—3

节点2—2

窗立面

盖瓦

底瓦

脊底瓦

沟头

滴水

脊盖瓦

望砖

檐口板

檐口大样

脊盖瓦
碎砖和土
脊底瓦
木板
檩条 φ10
垫木
望砖
椽子

盖瓦
底瓦
望砖
飞檐椽

望砖
椽子

檐部剖面

脊瓦
碎瓦和土
盖瓦
底瓦
望砖
椽子

垫木

横剖面

香洲细部大样图
Roof eaves, window, and tilework of Xiangzhou Pavilion

三十六鸳鸯馆
Hall of Thirty-six Mandarin Ducks

Hall of Thirty-six Mandarin Ducks is the central venue for staging activities in Humble Administrator's Garden's west section.

The hall is built in a square plan with an auxiliary service room in each corner. The space is divided into north and south sections by a central partition. The south section is designed for winter and spring, as it has a row of camellia trees in front—hence its original name "Hall of Eighteen Mandala Flowers". The north section is kept for spring and autumn use where one can view the water and distant hills; the pond nearby keeps mandarin ducks which hints the naming of the hall. The building's front and back are lined with floor-to-ceiling windows, weaving the inside and outside into a seamless whole.

三十六鸳鸯馆位于拙政园西部居中位置，是西部园林的主要厅堂和活动场所。

建筑主体平面呈正方形，四角各加一个耳室，作为服务用房。主厅内又用屏风划分为南北两半，南面宜于冬春，院前植有一排山茶花，故称『十八曼陀罗花馆』。北面宜于春秋，可远观山水景色，靠近还有水池，内养鸳鸯，故称『三十六鸳鸯馆』。建筑前后均采用长条落地窗，可以使内外空间融为一体。

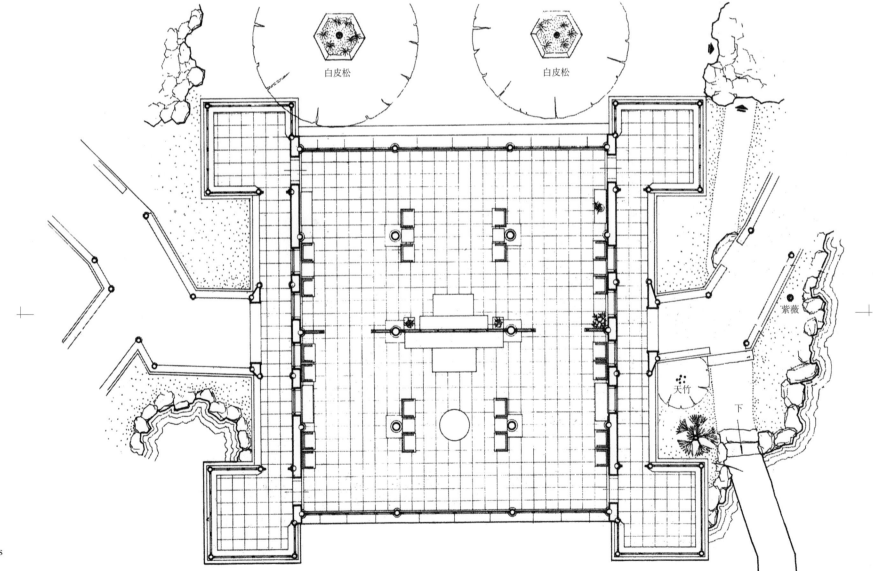

三十六鸳鸯馆平面图
Site plan of Hall of Thirty-six Mandarin Ducks

119

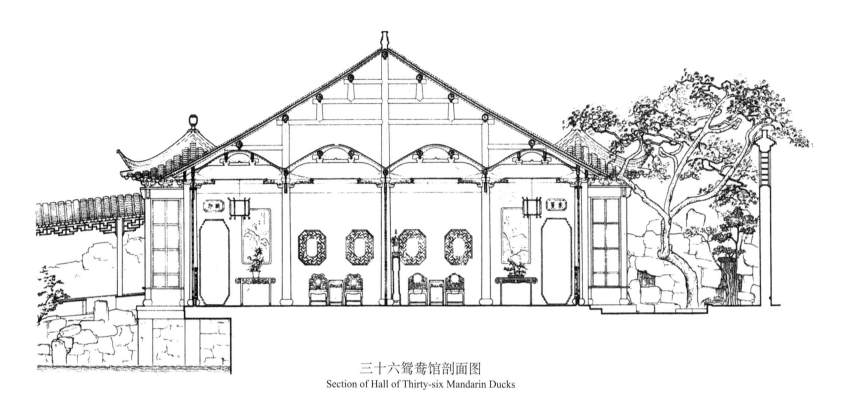

三十六鸳鸯馆剖面图
Section of Hall of Thirty-six Mandarin Ducks

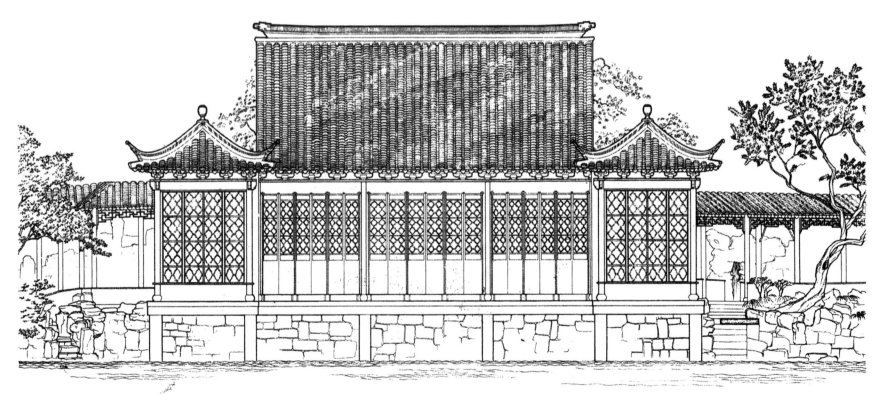

三十六鸳鸯馆北立面图
North elevation of Hall of Thirty-six Mandarin Ducks

0　1　2　3m

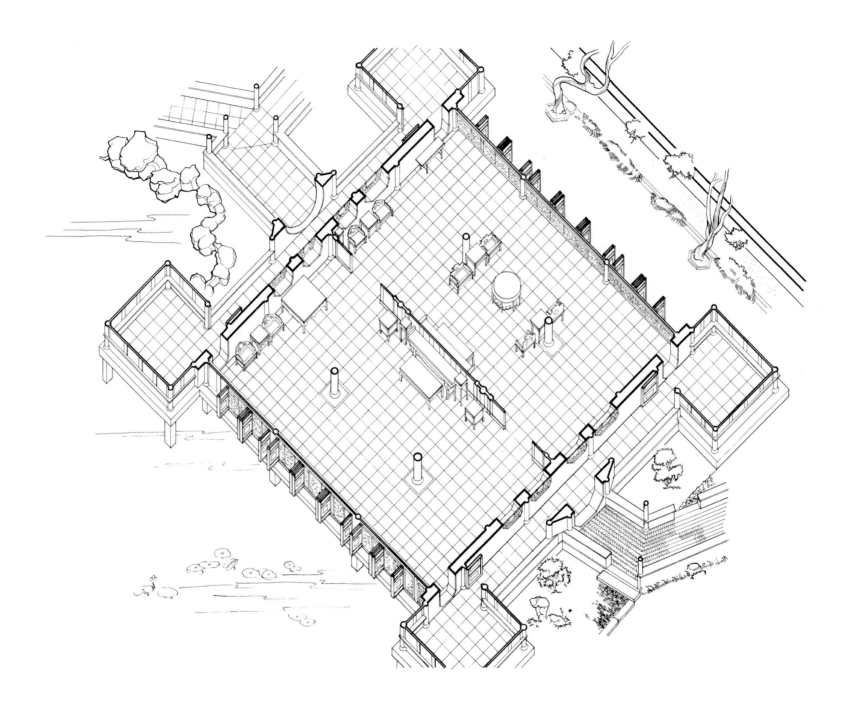

三十六鸳鸯馆透视图
Perspective of Hall of Thirty-six Mandarin Ducks

倚玉轩
Yiyuxuan

Yiyuxuan (Lean on Jade Pavilion) is a secondary, support building sitting west of Yuanxiang Hall.

The pavilion is built with a rectangular layout facing east and west within the immediate reach of the hall. Topped by a curved xieshan (hip-gable) roof, the interior is defined by tall openings on all sides—a desirable spot for viewing the garden. Meanwhile, Yiyuxuan shares the form and style of Yuanxiang Hall; when viewed from the north, their staggered profiles coincide into a cohesive view.

倚玉轩是拙政园主要厅堂远香堂的附属建筑，是一座次要的厅堂。

建筑平面长方形，主立面东西向，便于与远香堂往来。建筑外观为卷棚歇山式屋顶，四面均有长窗环绕，也是观赏园景最佳之处。同时，倚玉轩在外观上与远香堂构成一个群体，从北面向南望，纵横交错，景观极好。

紫薇

0 1 2 3m　N

倚玉轩平面图
Site plan of Yiyuxuan

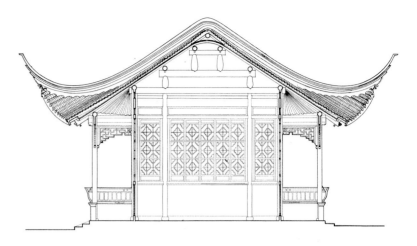

倚玉轩剖面图
Section of Yiyuxuan

倚玉轩西立面图
West elevation of Yiyuxuan

倚玉轩北立面图
North elevation of Yiyuxuan

0   1   2m

梧竹幽居亭
Wuzhu youju Pavilion

One can find the secluded, square-planned Wuzhu youju (Bamboo Cottage) Pavilion tucked in the north of Humble Administrator's Garden's midsection. Beneath a pyramidal roof with uplifted corners, the pavilion hosts a moon gate on each wall and is encircled by a columned gallery. Gazing west, Beisi (North Monastery) Pagoda comes into view, a "borrowed scene" of the garden; accompanied with the nearby lotus pond, it is one of the important spots for appreciating the sceneries in the garden.

梧竹幽居亭位于拙政园中部的东端，平面正方形。亭四面用圆洞门进出，外面有一圈柱廊，屋顶为攒尖顶，屋角为嫩戗发戗式，亭的西面可以远眺北寺塔为借景，近处有纵深的荷花池，是观赏园景的重要位置之一。

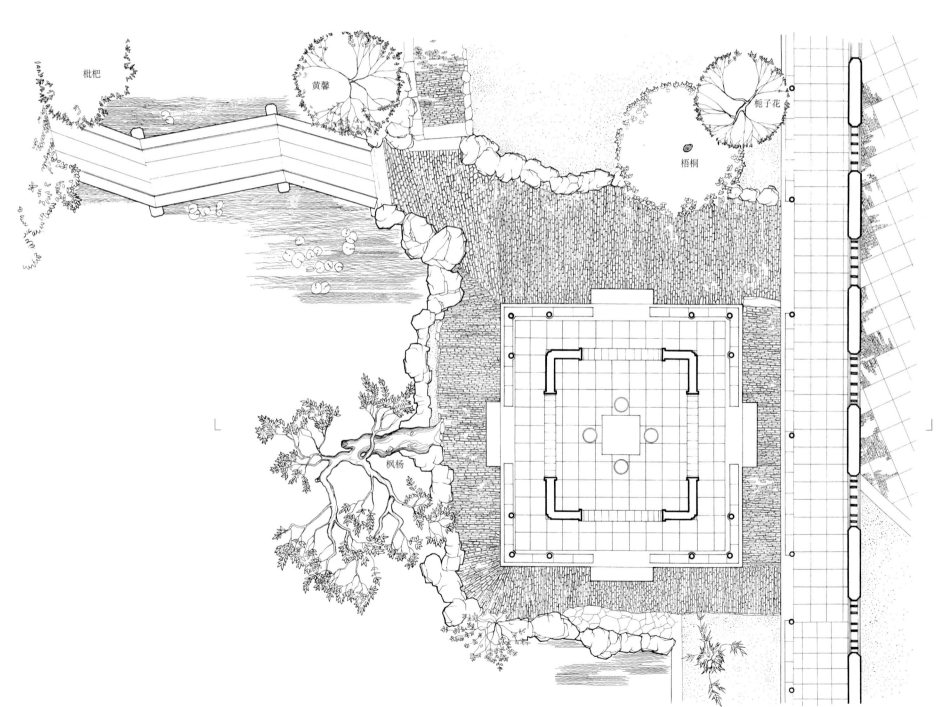

枇杷

黄馨

栀子花

梧桐

枫杨

N

0 1 2m

梧竹幽居亭平面图
Site plan of Wuzhu youju Pavilion

塔影亭剖面图
Section of Taying Pavilion

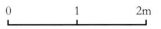

0　　　　1　　　　2m

塔影亭东立面图
East elevation of Taying Pavilion

0　　　1　　　2m

塔影亭南立面图
South elevation of Taying Pavilion

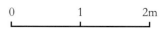

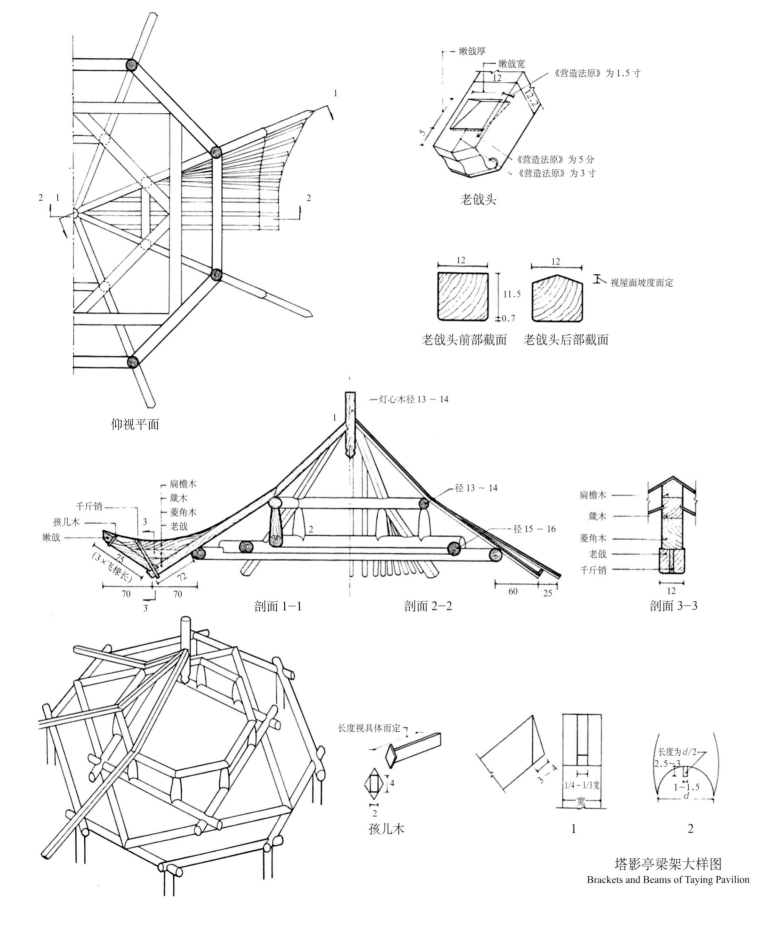

仰视平面

老戗头

嫩戗厚
嫩戗宽
《营造法原》为 1.5 寸
《营造法原》为 5 分
《营造法原》为 3 寸

老戗头前部截面　老戗头后部截面

视屋面坡度而定

一灯心木径 13～14

扁檐木
箴木
菱角木
老戗
千斤销
孩儿木
嫩戗

扁檐木
箴木
菱角木
老戗
千斤销

(3×飞椽长)

径 13～14
径 15～16

剖面 1-1　　　剖面 2-2　　　剖面 3-3

长度视具体而定

孩儿木

长度为 d/2
2.5～3
1～1.5

1/4～1/3宽
宽

1　　　　2

塔影亭梁架大样图
Brackets and Beams of Taying Pavilion

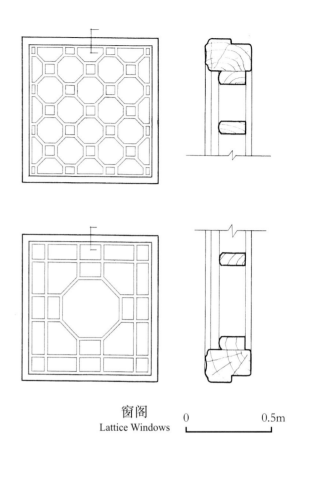

窗阁
Lattice Windows

0　　　0.5m

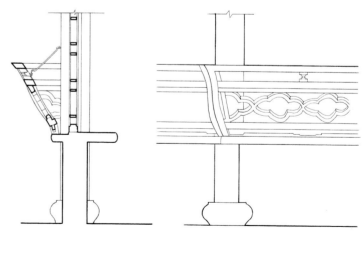

吴王靠
Wuwang Kao

0　　1　　2m

听雨轩庭院
Tingyuxuan courtyard

Tingyuxuan (Hearing Rain Pavilion) sits east inside Loquat Garden.

Rectangular in layout, the structure employs long windows in the front and rear. It is connected by a looping corridor, which leads to Haitang chunwu to the east and Linglong Pavilion to the west. To its south, a patch of Japanese banana trees reminds passersby of the sound of raindrops tapping against its leaves.

A waterlily pond sits in the middle of the garden, surrounded by *yulan* trees, sweet olive trees, Japanese banana trees and bamboo, staging a scene which varies by season. However, the pond's limited size and extruded rockery edge only furthered the confined feeling of the space.

听雨轩庭院在枇杷园之东。

听雨轩采用四面厅形式，前后用长窗。东面有长廊通海棠春坞，西面曲廊与玲珑馆相接。轩南满植芭蕉，取雨打芭蕉之意。轩北院中有水池，植睡莲，池周环植玉兰、桂花、芭蕉、竹丛等，四季景色变幻。但水池小、池岸高，显得局促逼仄。

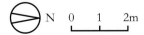

N   0   1   2m

听雨轩庭院平面图
Site plan of Tingyuxuan courtyard

听雨轩庭院 1-1 剖面图
Section 1-1 of Tingyuxuan courtyard

听雨轩庭院 2-2 剖面图
Section 2-2 of Tingyuxuan courtyard

0 1 2m

听雨轩庭院透视图

Perspective of Tingyuxuan courtyard

Haitang chunwu (Haitang Spring Pavilion) is situated north of Tingyu Pavilion within a small, secluded courtyard.

The two-bays wide building faces south, topped by yingshan (gable) roof; its west bay larger than its east. A gallery encircles the courtyard from east to west with a whitewashed wall to the south. Along the wall grows chuisi haitang (hall crabapple), the west with mugua haitang (Chinese quince). The courtyard's pebble pavement is also patterned to resemble haitang (Chinese crab apple) blossoms. Bamboos and lake rockeries are dotted throughout, accompanying the scene.

The pavilion was downsized due to limited space. Meanwhile, the gallery remains open and its decorations intricate. Although the scenes are scattered, the perforated openings, moon gates, and windows ties them together by framing objects from a variety of angles. Although the courtyard is small, the treatment of its component enhanced its layout, making it a fine example in small garden design.

海棠春坞庭院
Haitang chunwu Courtyard

『海棠春坞』在听雨轩之北，是一组精巧幽静的小院落。

主体建筑坐北朝南，为两开间硬山房屋，东间小，西间大，形制自由。庭院东西两面均以走廊相连，南面为粉墙。庭院中植垂丝海棠，轩西小院植木瓜海棠，地面也用卵石铺成海棠纹图案。此外再用少许竹丛、湖石作为陪衬。

此院面积很小，建筑物尺度相应减小，房屋走廊空透，装修细部精巧秀丽，景物稀疏，漏窗、洞门、空窗富于变化，格外显得小巧玲珑，是小庭院处理的精品。

海棠春坞庭院平面图
Site plan of Haitang chunwu Courtyard

海棠春坞 1-1 剖面图
Section 1-1 of Haitang chunwu Courtyard

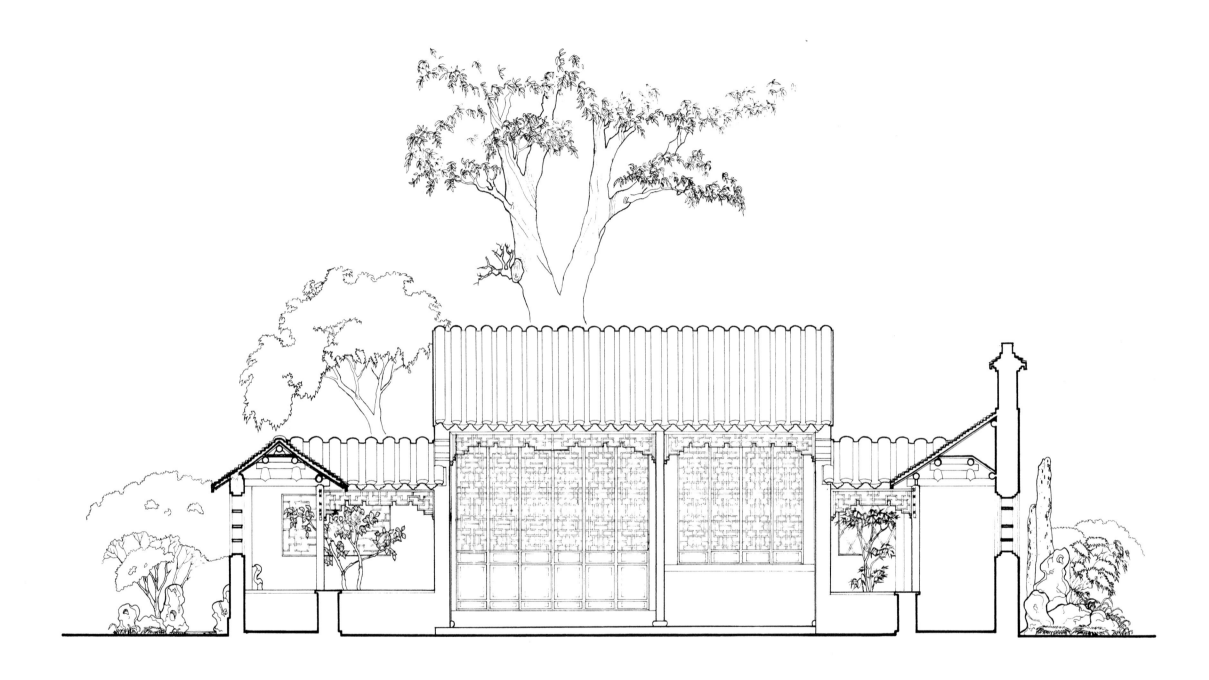

海棠春坞 2-2 剖面图
Section 2-2 of Haitang chunwu Courtyard

0　　1　　2m

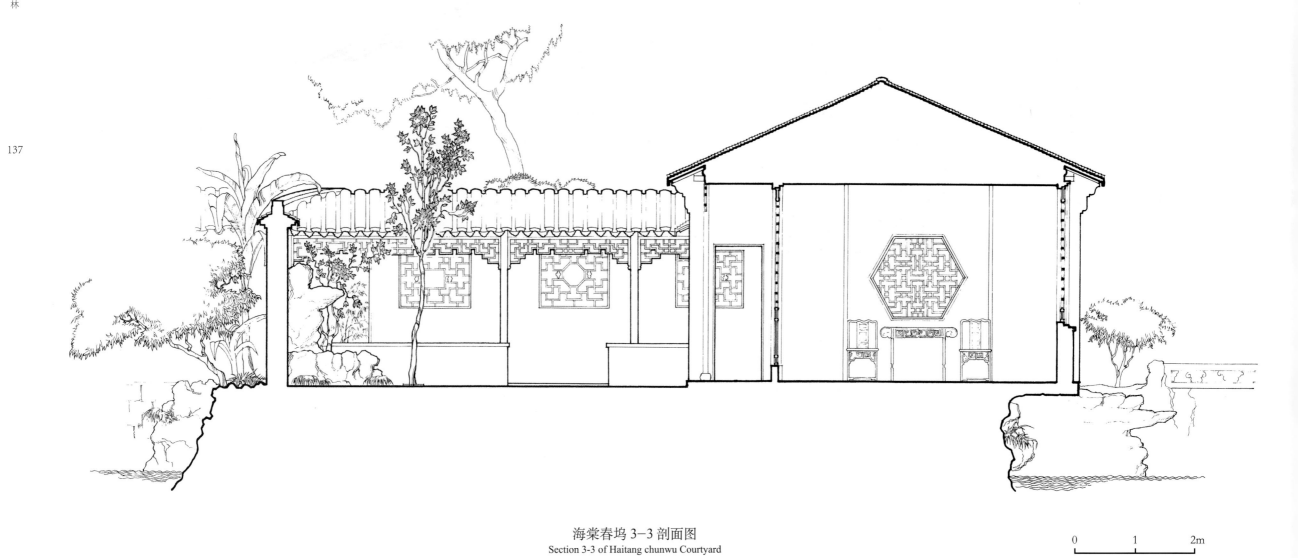

海棠春坞 3-3 剖面图
Section 3-3 of Haitang chunwu Courtyard

0　　　1　　　2m

海棠春坞透视图
Perspective of Haitang chunwu Courtyard

小沧浪水院
Xiaocanglang
Water Courtyard

Xiaocanglang (Small Surging Wave) water courtyard sits west of Humble Administrator's Garden's center gate.

Xiaocanglang Water Pavilion, the main structure, is elevated above the pond, facing Xiaofeihong (Small Rainbow) Bridge; viewing northward from here, the distant scenery appears deep and layered. The bridge, set amidst shaded greenery, perches over the emerald, shimmering water in a graceful arc. To its west stands Dezhen (Truth) Pavilion; to the east, Song feng (Hearing Rustling Pine) Pavilion—their collective presence adding to an otherwise monotonous space. To the south of Dezhen Pavilion, a particularly narrow, walled gallery is set in stark contrast to the open pavilions, galleries and the bridge.

The water courtyard may appear complex, yet its treatments are minimal: its spaces are partitioned yet connected, merging inside and outside into one. The water also extends beneath the pavilions, appearing to flow on endlessly. Such nuanced techniques are what created the microcosmic scene within; though limited in space, the water imparted the imagery of a Jiangnan village. It is an exemplary water courtyard design.

小沧浪水院在拙政园腰门之西，以水景为主。

小沧浪水阁为主建筑，架临在水面上，正北直对小飞虹廊桥，由此北望，景色深远，层次较多。绿荫虹桥，碧波游鳞，环境清幽。水阁西有得真亭，东置松风亭，打破构图的单调。得真亭南一段走廊的墙面与周围透空的亭阁廊桥形成虚实对比。

建筑处理虚多实少，空间隔而不分，内外贯通渗透连成一片；房屋与水景结合紧密，水面延伸于小沧浪与松风亭下；环境小却有小中见大之妙，水面不大而有水乡弥漫之意，为水院的典型作品。

① 小沧浪 Xiaocanglang Water Pavilion
② 松风亭 Songfeng Pavilion
③ 得真亭 Dezhen (Truth) Pavilion
④ 小飞虹廊桥 Xiaofeihong (Small Rainbow) Bridge

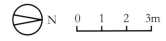

N   0 1 2 3m

小沧浪水院平面图
Site plan of Xiaocanglang Water Courtyard

小沧浪水院透视图
Perspective of Xiaocanglang Water Courtyard

小沧浪水院透视图
Perspective of Xiaocanglang Water Courtyard

小沧浪水院剖面图
Section of Xiaocanglang Water Courtyard

0　1　2　3m

见山楼
Jianshan Building

Jianshan (Mountain View) Building stands in the northwestern corner of Humble Administrator's Garden's midsection; it is a two-storied building topped by a double-eaves *xieshan* (hip-gable) roof. As a main building of the area, its ground floor is a large, open room used for daily activities. Despite its size, there are no stairs inside; instead, one can only reaches its second floor via an external gallery.

见山楼位于拙政园中部的西北隅，是一座平缓的二层建筑，外观为卷棚歇山式顶。底层为日常活动之用，内部空间开阔，体形较大，是这一片区的主体建筑。楼内无楼梯，上层通过外部的爬山廊进入。

榆树

松树

N

0 1 2 3m

见山楼一层平面图
Plan of first floor of Jianshan Building

见山楼二层平面图
Plan of second floor of Jianshan Building

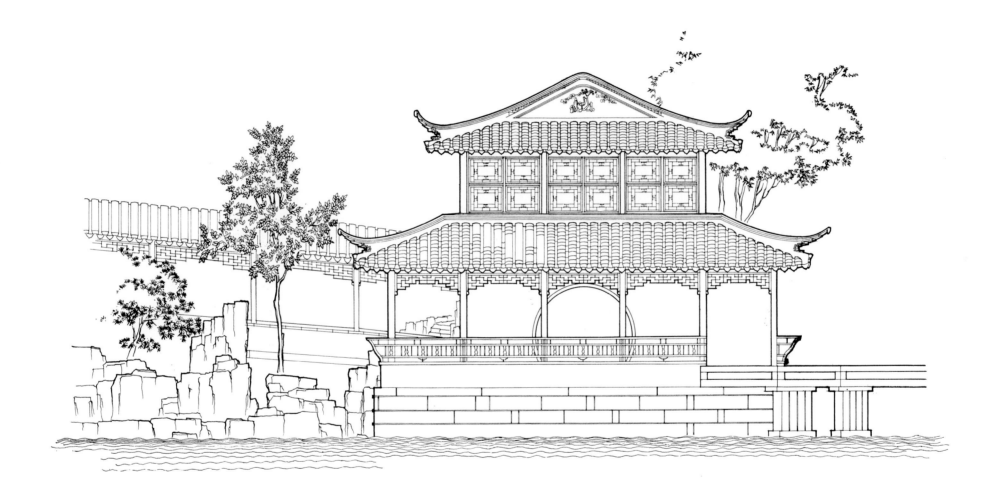

见山楼东立面图
East elevation of Jianshan Building

0　　1　　2　　3m

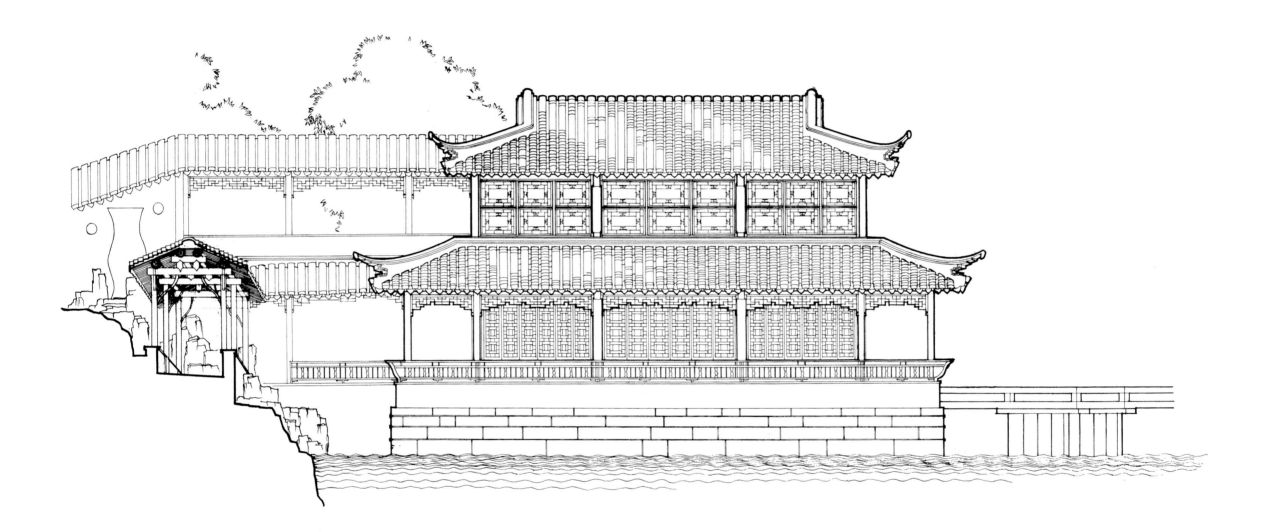

见山楼南立面图

South elevation of Jianshan Building

0   1   2   3m

雪香云蔚亭
Xuexiang yunwei Pavilion

Xuexiang yunwei (Fragrant Snow, Splendid Cloud) Pavilion sits on a small hill of the isle in Humble Administrator's Garden's midsection. It is the corresponding scene of Yuanxiang Hall when viewed northward.

Rectilinear in plan, the pavilion has a hip-gable roof with uplifted eaves. Surrounded by plum blossoms and wintersweets, the blossoms depart a memorable fragrance upon winter, hence its name Xuexiang.

Supported by granite pillars, the hilltop pavilion naturally blends into its setting. Inscribed on two pillars, a couplet reads "a cicada's noise softens the forest, a birds' chirp fades away the mountain," adding a poetic touch to the scene.

雪香云蔚亭位于拙政园中部的小山上，是主厅远香堂的北面对景。

建筑平面长方形，屋顶为卷棚歇山，屋角水戗发戗，纯朴自然。四周种梅花、蜡梅，冬季闻香之处，是为『雪香』。

该亭位于山上，故柱子采用花岗石建造，与周围环境融为一体。亭前有对联一副：『蝉噪林愈静，鸟鸣山更幽』，点出诗情画意的内涵。

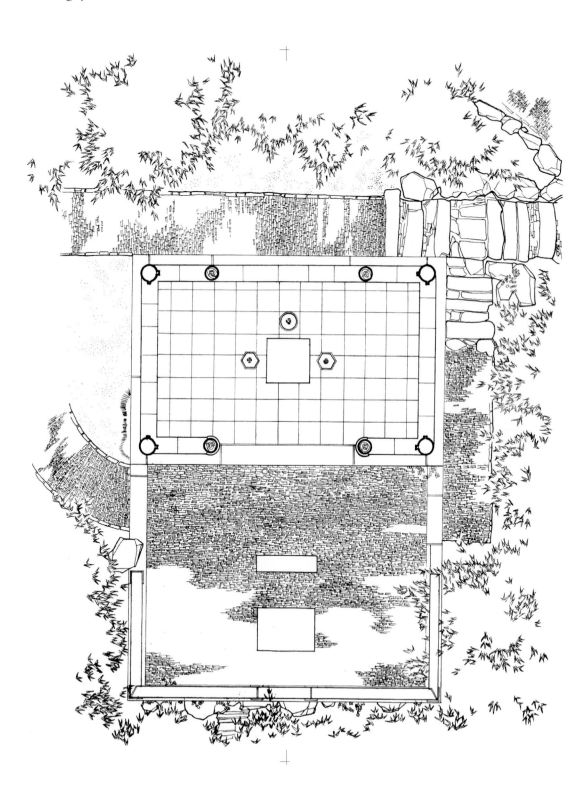

雪香云蔚亭平面图
Site plan of Xuexiang yunwei Pavilion

雪香云蔚亭剖面图
Section of Xuexiang yunwei Pavilion

雪香云蔚亭南立面图
South elevation of Xuexiang yunwei Pavilion

0　　　1　　　2m

荷风四面亭
Hefeng simian Pavilion

Hefeng simian (Amidst Lotus) Pavilion occupies a prominent position in Humble Administrator's Garden's midsection. Hexagonal in plan, the pavilion opens on all sides. A couplet hanging before the columns reads "Four walls of lotus, three sides of willow, half a pond of autumn water—set amidst a hill".

荷风四面亭位于拙政园中部的主要位置。此处人流集中，位置显要。亭作六角形，周围开敞，柱前挂有一副对联：『四壁荷花三面柳，半潭秋水一房山』。

乌柏

柳树

柏树

柏树

N 0 1 2 3m

荷风四面亭平面图
Site plan of Hefeng simian Pavilion

荷风四面亭立面图
Elevation of Hefeng simian Pavilion

0　　　　1　　　　2m

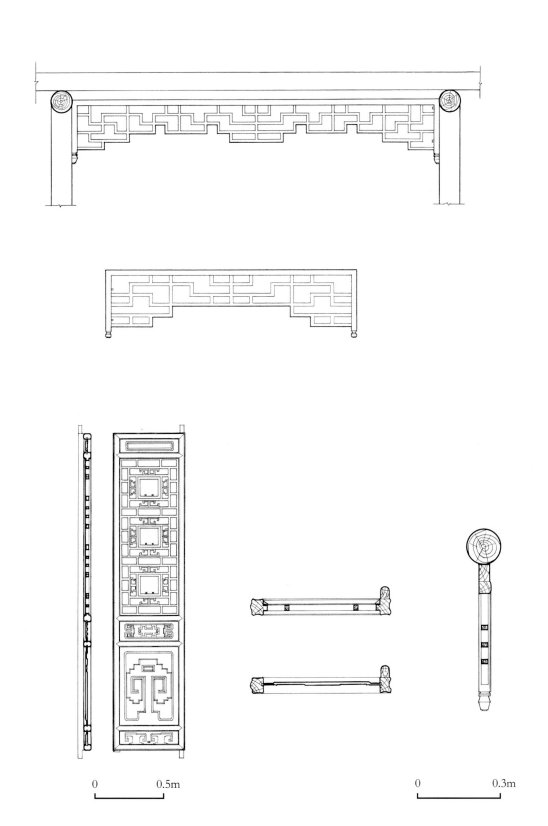

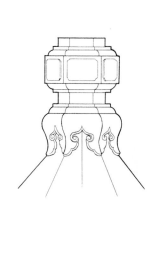

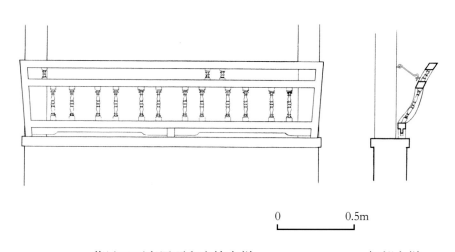

见山楼内檐装修
Interior decoration and partition of Jianshan Building

荷风四面亭屋顶和座椅大样
Roof and seating of Hefeng simian Pavilion

细部大样
Construction details

Yipu Suzhou

Built in Ming dynasty, Yipu (Garden of Cultivation) is located in No. 5 Wenya Lane, Suzhou. Covering an area of 0.33 ha, the garden has kept its original layout since the late Ming and early Qing dynasties.

The garden is centered on the pond of 670 m sq; inlets flow from its southeast and southwest corners, each topped by a stone footbridge. The north of the pond is marked by numerous architecture, including a five-bays wide Waterside Pavilion. Perched above the pond, this pavilion is flanked by side rooms to its east and west. To the pond's south stands the garden's main scene: an earthen and rockery hill with steep cliff faces and narrow detours tucked along the bank. Looking from the north, it appears as if a lush forest grew out of a mountain. The eastern and western banks are traced by galleries and pavilions, accompanied by trees and rocks—offering open views and foliage as one walks through the garden.

Heading south along the pond's eastern edge, one arrives at Ruyu (Small Fish) Pavilion. Its unique wooden structure is a relic from Ming dynasty.

苏州艺圃

艺圃在苏州文衙弄五号，始建于明。现山池布局大致因明末清初旧况，面积约0.33公顷。

园的总体布局以水池为中心，水池占地约670平方米，布置以聚为主，东南角和西南角伸出水湾各一，水口各架石板桥一座。池北以建筑为主，建水榭五间，挑于池面，水榭两侧厢房也临池而和池东、西厢房相连属。池南堆土叠石为山，临池用湖石叠成绝壁及危径，从池北眺望，山石嶙峋，树木葱郁，是园中的主要对景。池东、西两岸以疏朗的亭廊、树石作为南、北之间的过渡与陪衬。

由水榭东厢房折南，沿东岸小径至乳鱼亭，此亭木构系明代遗物。

① 门厅 Entrance Hall
② 思嗜轩 Sishixuan
③ 乳鱼亭 Ruyu Pavilion
④ 延光阁（水榭）Yanguangge
⑤ 博雅堂 Boya Hall
⑥ 爱莲窝 Ailianwo
⑦ 世伦堂 Shiluntang
⑧ 东莱草堂 Donglai caotang
⑨ 馎饦斋 Botuo Hall
⑩ 思敬居（绣楼）Sijingju
⑪ 七襄公所 Qixiang gongsuo
⑫ 响月廊 Xiangyue Gallery
⑬ 芹芦 Qinlu
⑭ 朝爽亭 Zhaoshuangting
⑮ 旸谷书屋 Yanggu shuwu
⑯ 敬亭山房 Jingting shangfang
⑰ 香草居 Xiangcaoju
⑱ 南斋 Nanzhai
⑲ 折桥 Curved Bridge
⑳ 直桥 Straight Bridge

苏州艺圃总平面图
Site plan of Yipu Suzhou

N 0 5 10m

苏州艺圃剖面图
Section of Yipu Suzhou

0　1　2　3m

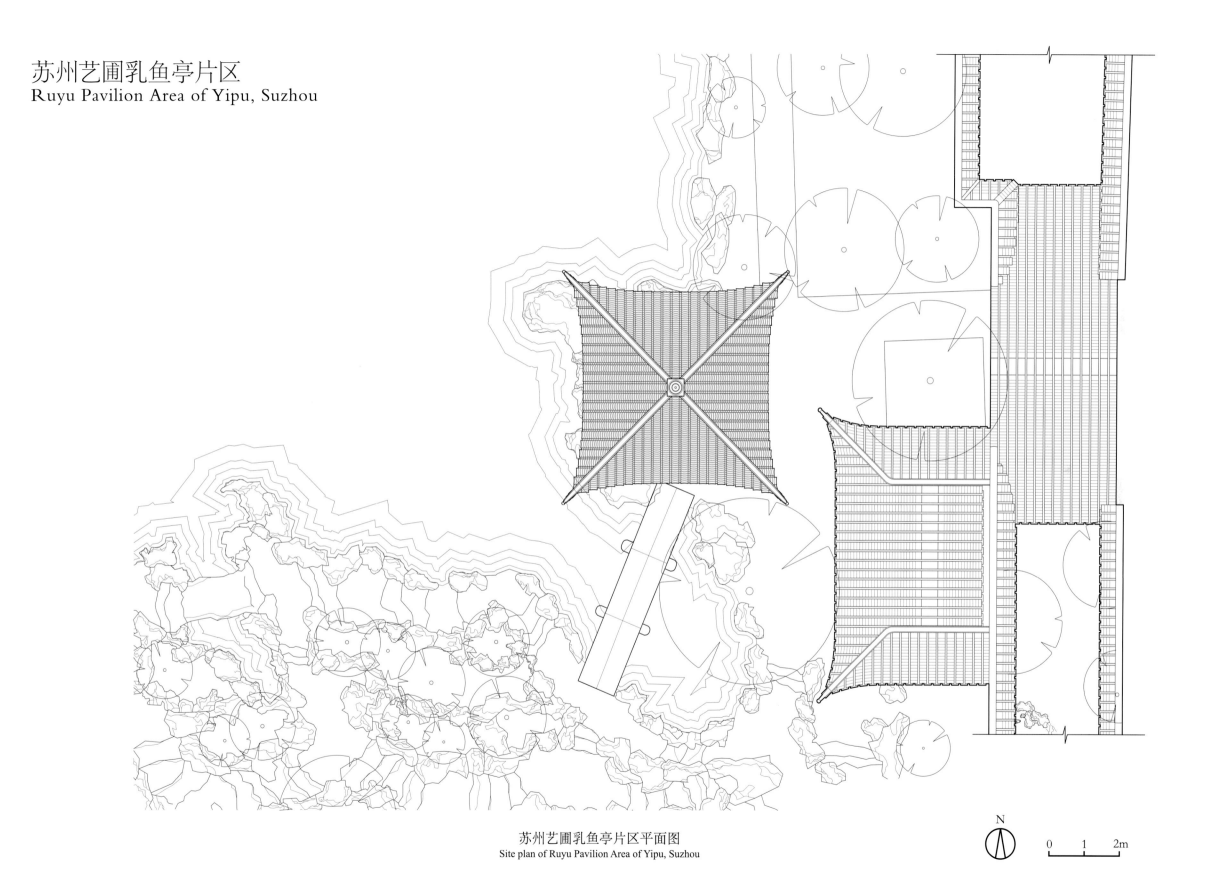

苏州艺圃乳鱼亭片区
Ruyu Pavilion Area of Yipu, Suzhou

苏州艺圃乳鱼亭片区平面图
Site plan of Ruyu Pavilion Area of Yipu, Suzhou

N

0   1   2m

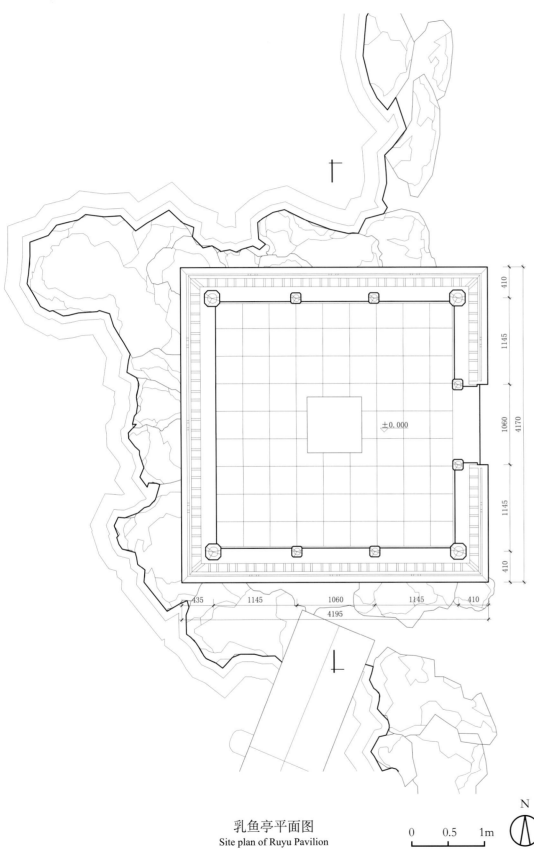

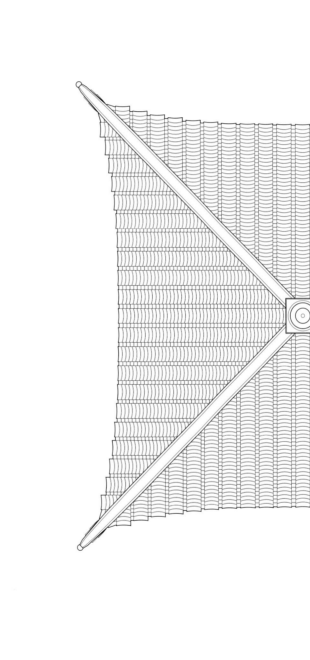

乳鱼亭平面图
Site plan of Ruyu Pavilion

0   0.5   1m

N

乳鱼亭屋顶平面图
Roof plan of Ruyu Pavilion

乳鱼亭仰视平面图
Reflected ceiling plan of Ruyu Pavilion

乳鱼亭剖面图
Section of Ruyu Pavilion

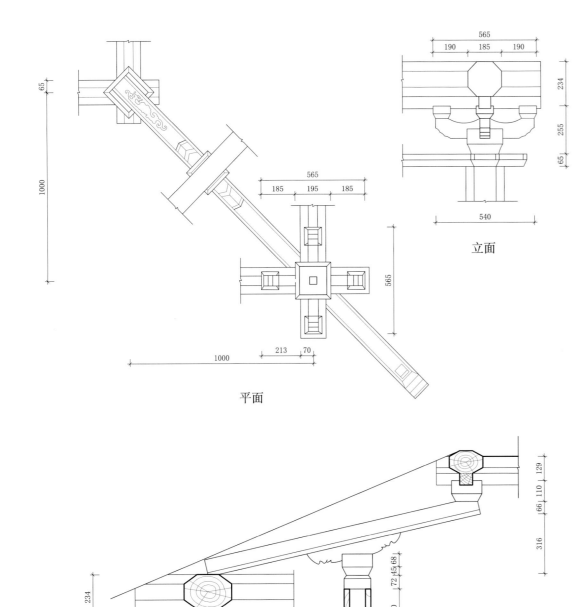

立面

平面

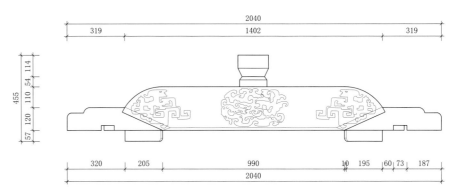

乳鱼亭抹角梁大样图
Oblique beam of Ruyu Pavilion

剖面

乳鱼亭转角斗拱大样图
Corner bracketing of Ruyu Pavilion

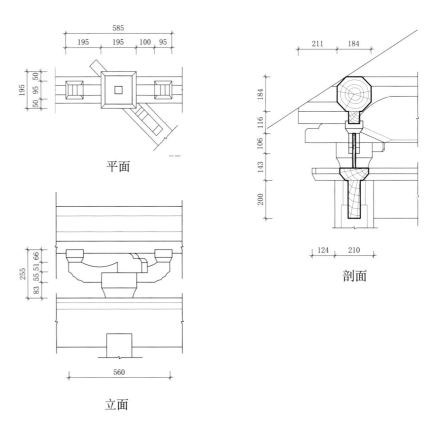

平面

剖面

立面

乳鱼亭柱头斗拱大样图
Column-top bracket set of Ruyu Pavilion

0   0.5m

苏州艺圃延光阁和博雅堂
Yanguangge and Boya Hall of Yipu, Suzhou

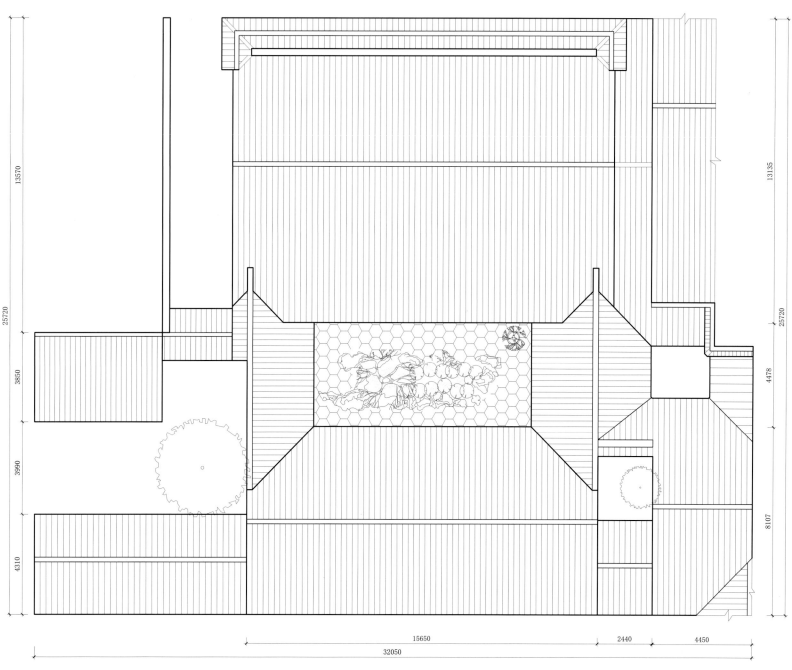

① 延光阁（水榭）Yanguangge
② 博雅堂 Boya Hall
③ 院子 courtyard
④ 敬亭山房 Jingting shangfang
⑤ 旸谷书屋 Yanggu shuwu

苏州艺圃延光阁和博雅堂平面图
Site plan of Yanguangge and Boya Hall of Yipu, Suzhou

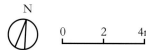

0   2   4m

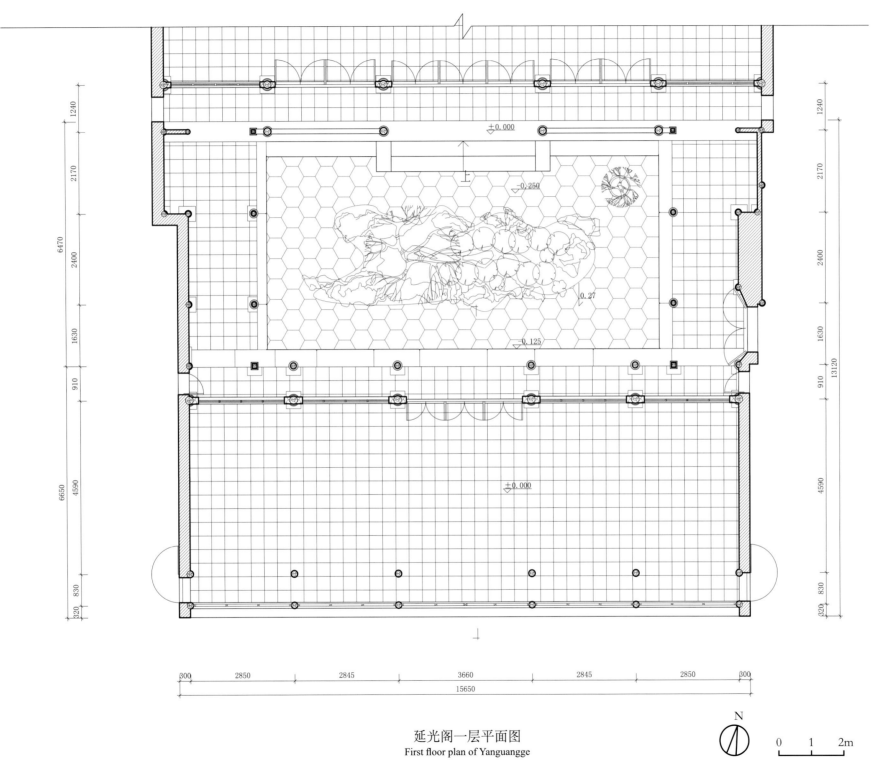

延光阁一层平面图
First floor plan of Yanguangge

N

0 1 2m

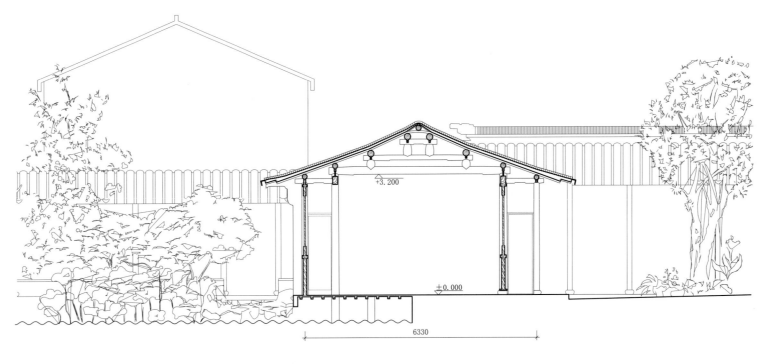

延光阁剖面图
Section of Yanguangge

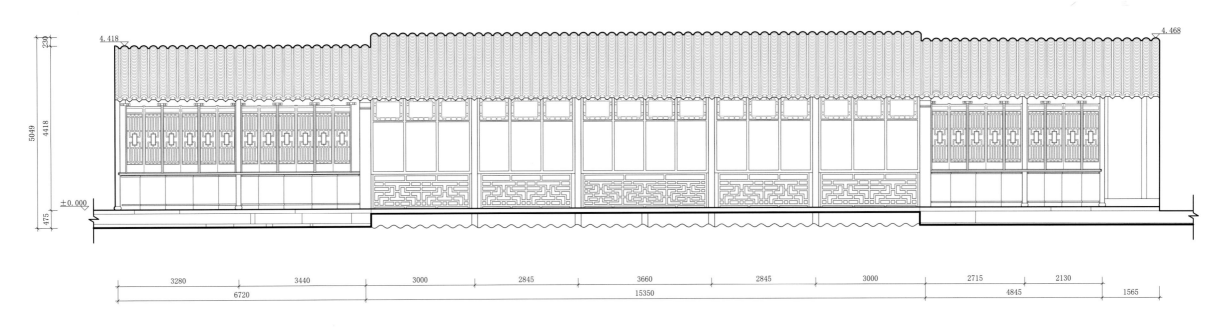

延光阁南立面图
South elevation of Yanguangge

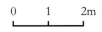

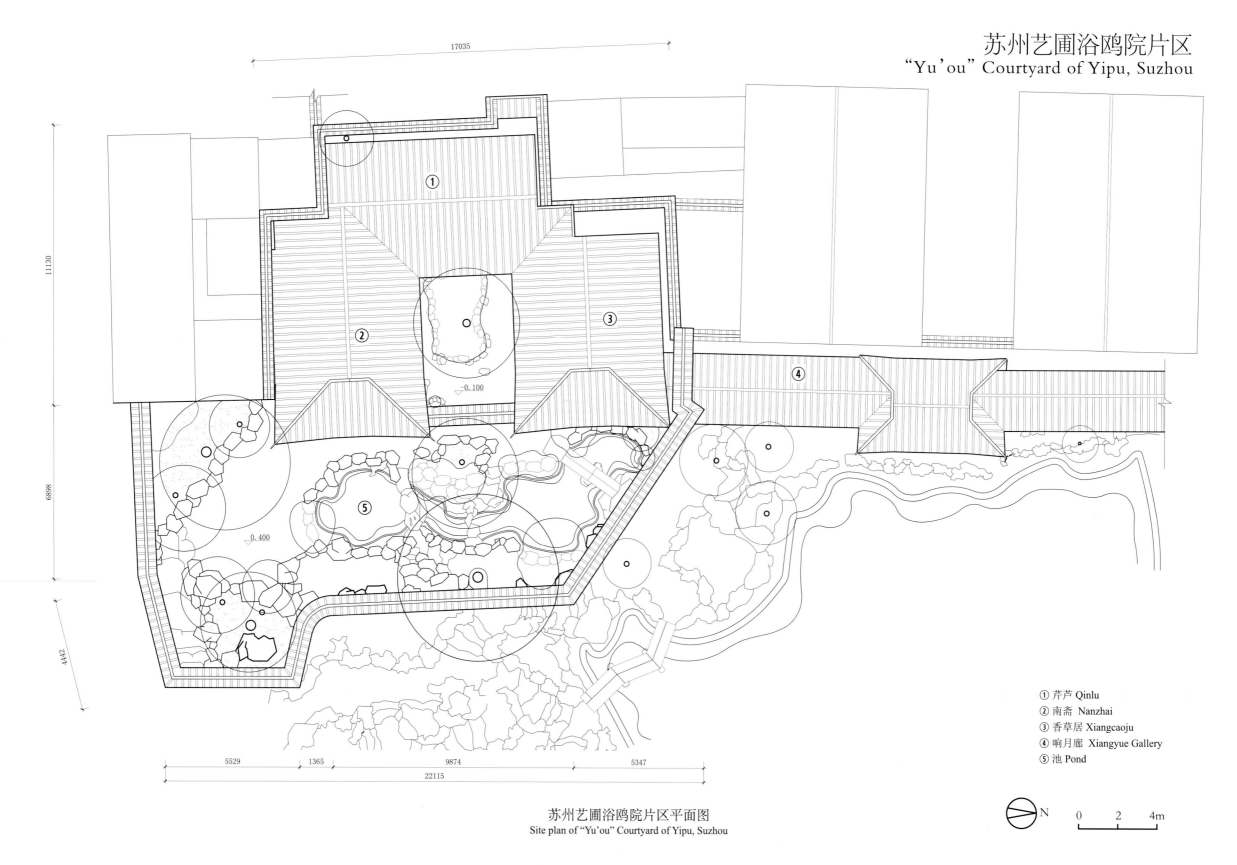

苏州艺圃浴鸥院片区
"Yu'ou" Courtyard of Yipu, Suzhou

① 芹芦 Qinlu
② 南斋 Nanzhai
③ 香草居 Xiangcaoju
④ 响月廊 Xiangyue Gallery
⑤ 池 Pond

苏州艺圃浴鸥院片区平面图
Site plan of "Yu'ou" Courtyard of Yipu, Suzhou

N   0   2   4m

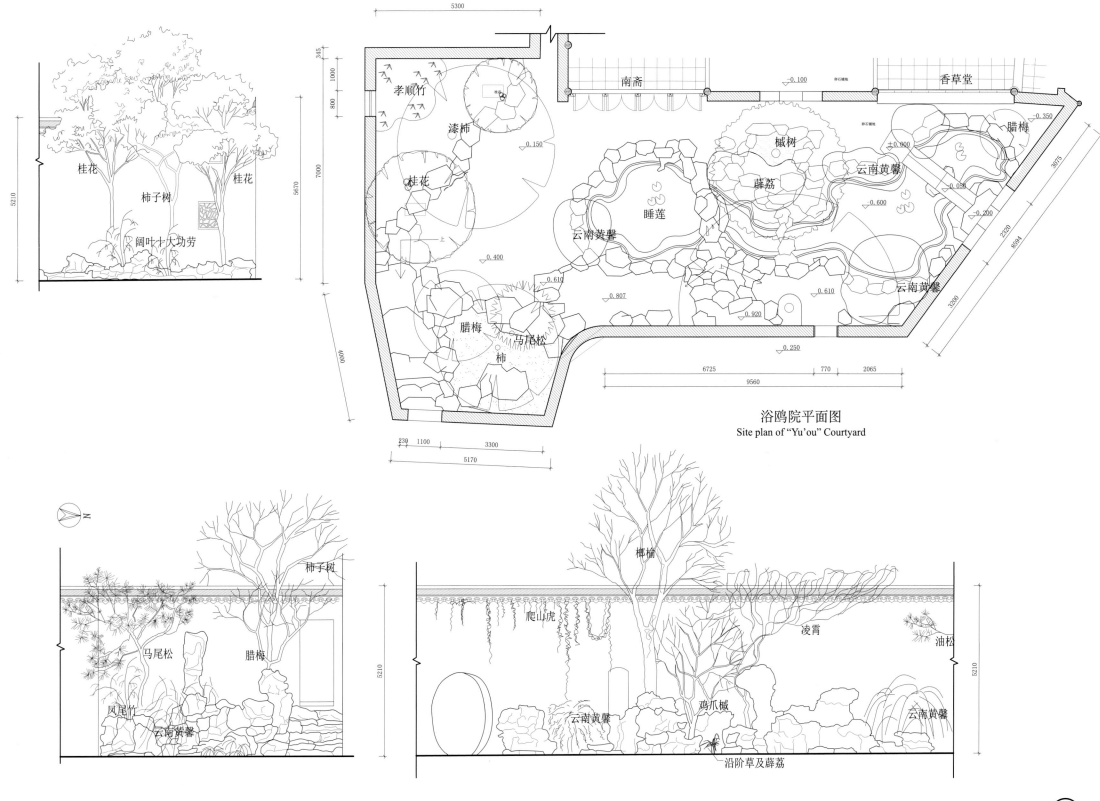

桂花
柿子树
桂花
阔叶十大功劳

孝顺竹
漆柿
桂花
腊梅
马尾松
柿
南斋
香草堂
腊梅
栖树
薜荔
云南黄馨
睡莲
云南黄馨
云南黄馨

浴鸥院平面图
Site plan of "Yu'ou" Courtyard

马尾松
腊梅
柿子树
凤尾竹
云南黄馨

榔榆
爬山虎
凌霄
油松
云南黄馨
鸡爪槭
云南黄馨
沿阶草及薜荔

浴鸥院东立面景观展开图
East elevation of "Yu'ou" Courtyard

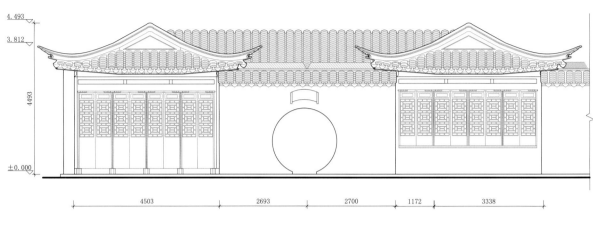

芹芦东立面图
East elevation of Qinlu

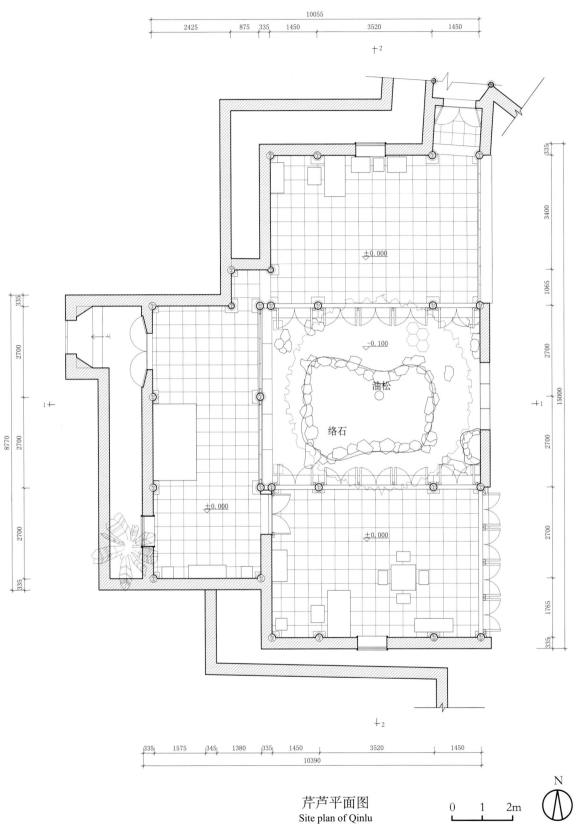

芹芦平面图
Site plan of Qinlu

0　1　2m

N

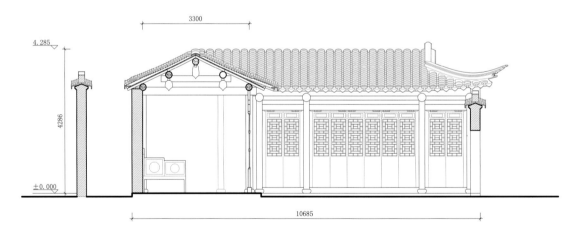

芹芦 1-1 剖面图
Section 1-1 of Qinlu

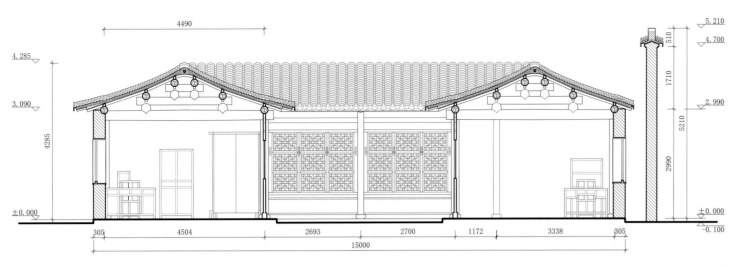

芹芦 2-2 剖面图
Section 2-2 of Qinlu

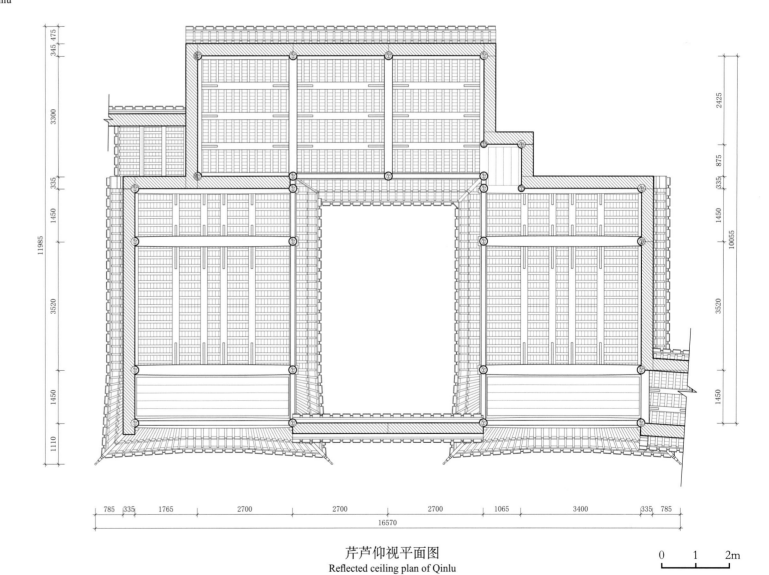

芹芦仰视平面图
Reflected ceiling plan of Qinlu

0 1 2m

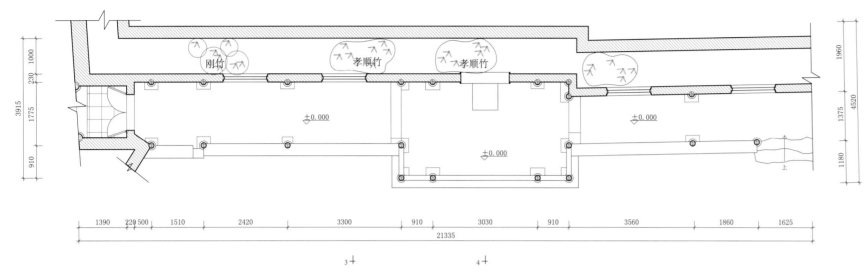

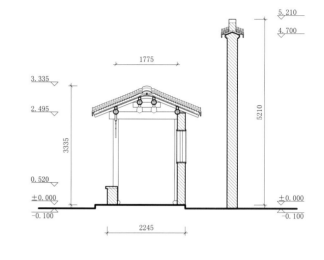

响月廊平面图
Site plan of Xiangyue Gallery

响月廊 3-3 剖面图
Section 3-3 of Xiangyue Gallery

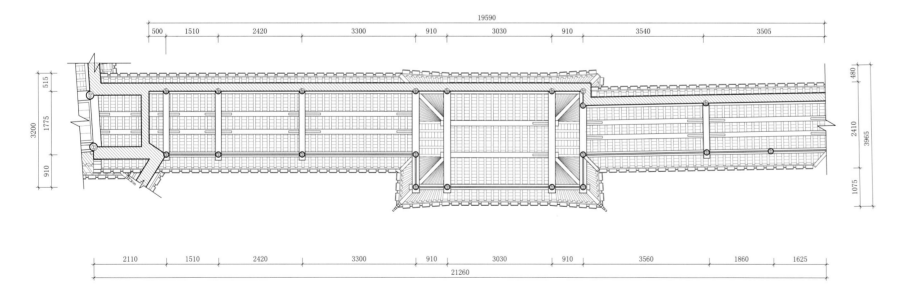

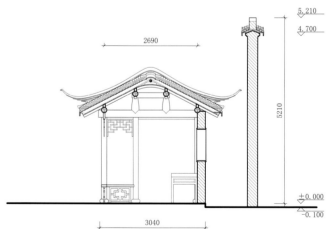

响月廊仰视平面图
Reflected ceiling plan of Xiangyue Gallery

响月廊 4-4 剖面图
Section 4-4 of Xiangyue Gallery

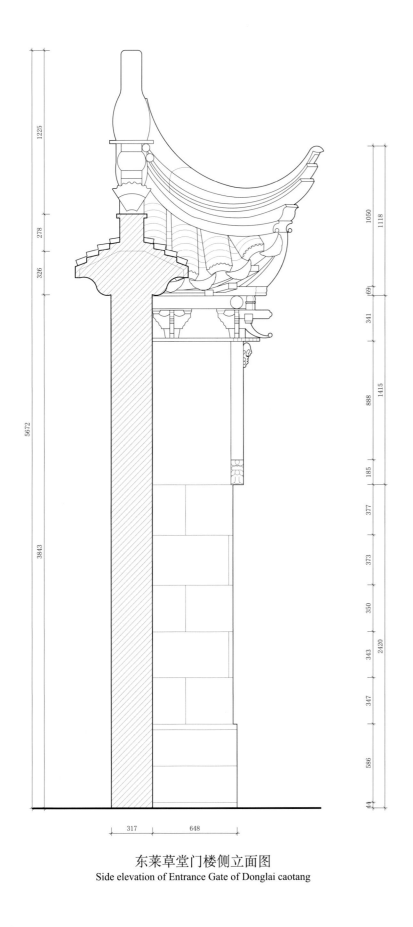

东莱草堂门楼侧立面图
Side elevation of Entrance Gate of Donglai caotang

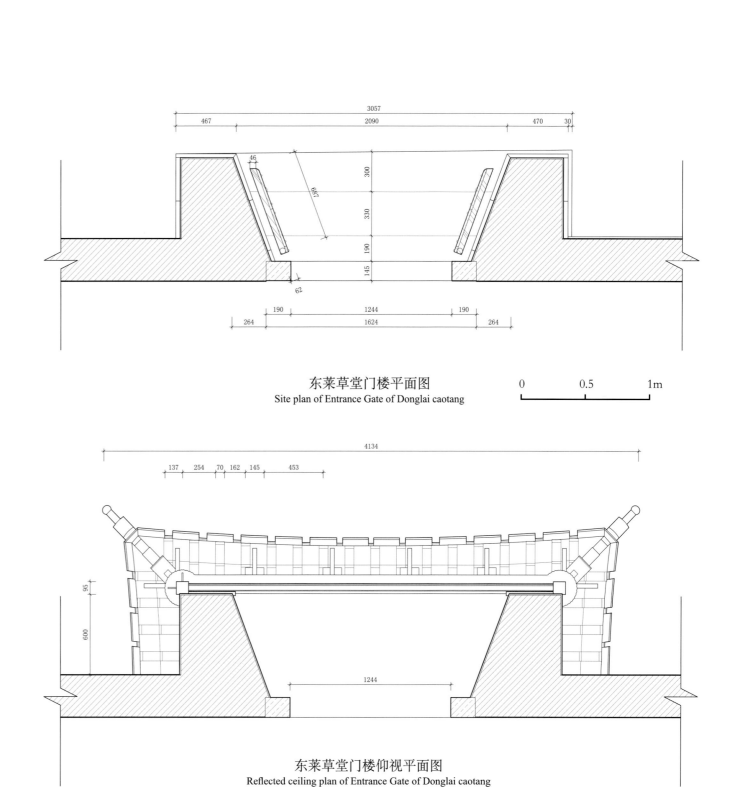

东莱草堂门楼平面图
Site plan of Entrance Gate of Donglai caotang

0　　　　0.5　　　　1m

东莱草堂门楼仰视平面图
Reflected ceiling plan of Entrance Gate of Donglai caotang

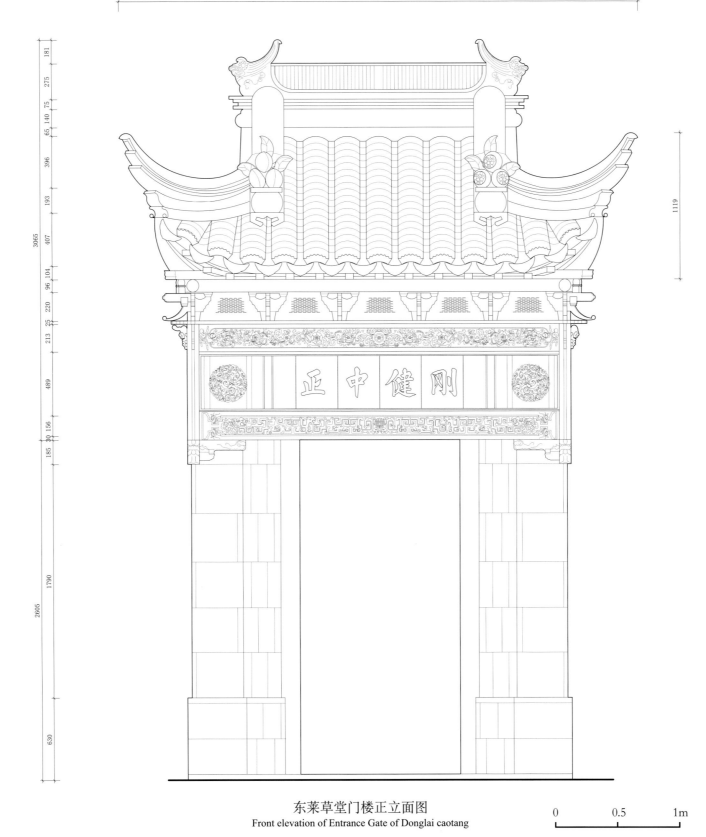

东莱草堂门楼正立面图
Front elevation of Entrance Gate of Donglai caotang

0   0.5   1m

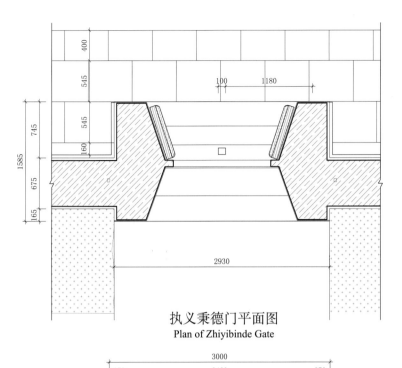

执义秉德门平面图
Plan of Zhiyibinde Gate

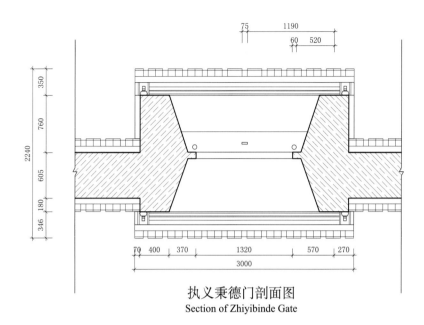

执义秉德门剖面图
Section of Zhiyibinde Gate

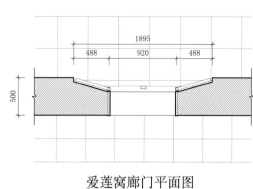

爱莲窝廊门平面图
Plan of door in Ailianwo Gallery

爱莲窝廊门仰视平面图
Reflected ceiling plan of door in Ailianwo Gallery

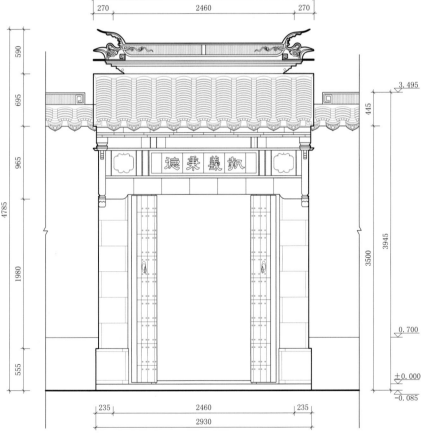

执义秉德门立面图
Elevation of Zhiyibinde Gate

0   0.5   1m

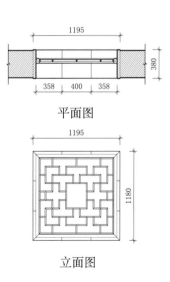

平面图

立面图

爱莲窝花窗细部
Lattice of windows in Ailianwo Gallery

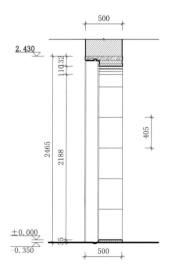

爱莲窝廊门剖面图
Section of door in Ailianwo Gallery

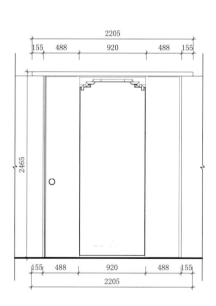

爱莲窝廊门南立面图
South elevation of door in Ailianwo Gallery

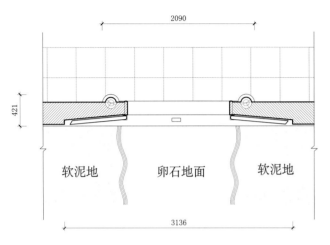

2090

421

软泥地　　卵石地面　　软泥地

3136

七襄公所西门平面图
Plan of west entrance in Qixiang gongsuo

N

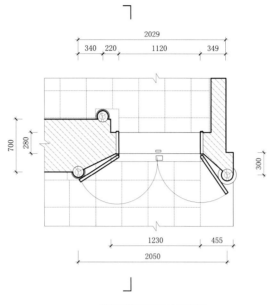

2029

340　220　1120　349

700

280

300

1230　455

2050

延光阁院门平面图
Plan of Gate of Yanguangge

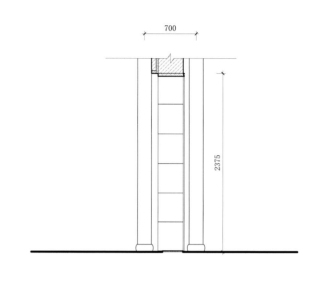

700

2375

延光阁院门剖面图
Section of Gate of Yanguangge

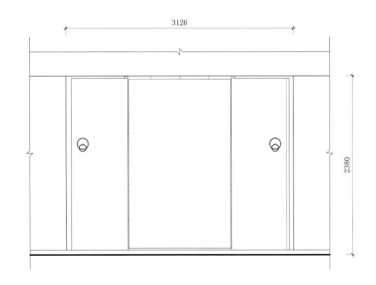

3126

2380

七襄公所西门南立面图
South elevation of west entrance in Qixiang gongsuo

0　　0.5　　1m

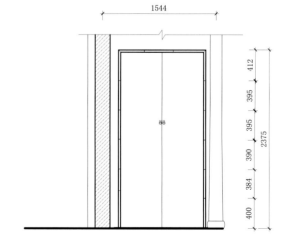

2050

475　645　598　332

2390

延光阁院门西立面图
West elevation of Gate of Yanguangge

1544

412
395
395
390　2375
384
400

延光阁院门东立面图
East elevation of Gate of Yanguangge

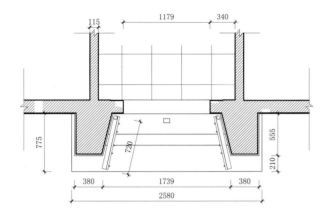

博饤斋院门平面图
Site plan of Gate of Botuo Hall

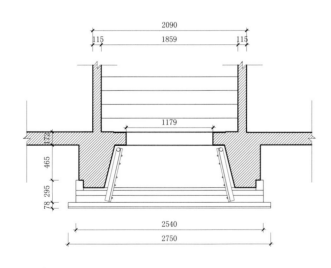

博饤斋院门仰视平面图
Reflected ceiling plan of Gate of Botuo Hall

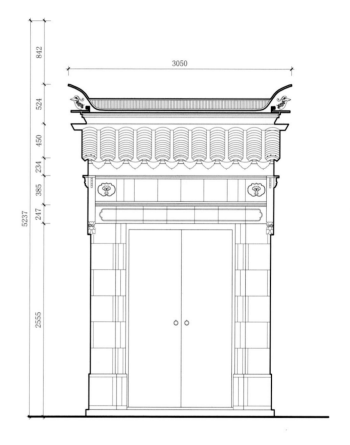

博饤斋院门剖面图
Section of Gate of Botuo Hall

博饤斋院门立面图
Elevation of Gate of Botuo Hall

0   0.5   1m

青砖铺地－方砖
Brick pavement—square tiles

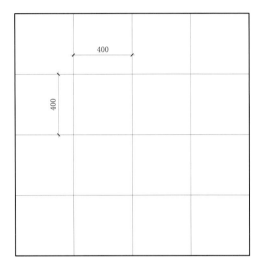

400×400 方砖
400mm×400mm tile

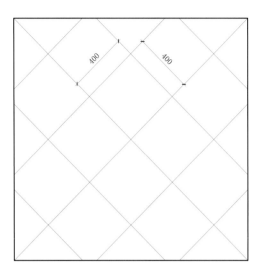

400×400 方砖（斜拼）
400mm × 400mm tile (oblique)

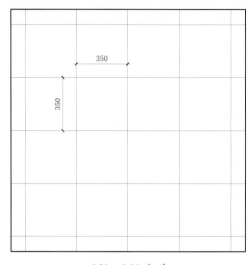

350×350 方砖
350mm×350mm tile

石材铺地
Stone pavement

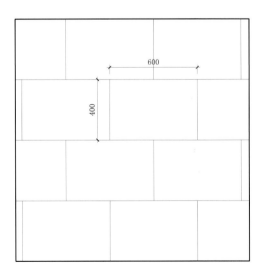

600×400 石材铺地－地坪石
600mm × 400mm tile—slab

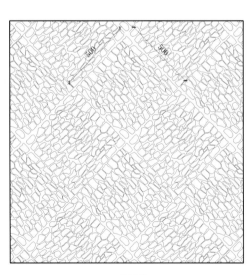

碎拼
Fragmented pavement

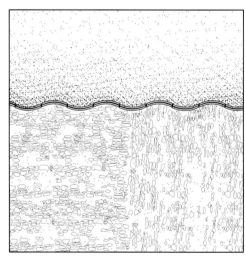

卵石铺地与软地交接
Junction between pavement and soil

苏州艺圃铺地大样图
Pavement Patterns of Yipu, Suzhou

0   0.5   1m

花街铺地－硬景
Floral pattern pavement - hardscape

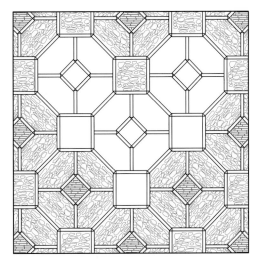

八角灯景
Octagonal lantern

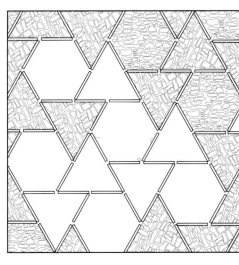

六角冰纹
Hexagonal ice pattern

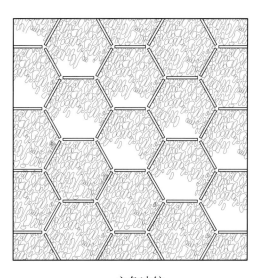

六角冰纹
Hexagonal ice pattern

青砖铺地－皇道砖
Brick pavement—royal brick

人字
Ren (people) pattern

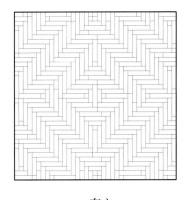

套方
Concentric pattern

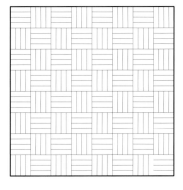

间方
Orthogonal pattern

苏州艺圃铺地大样图
Pavement Patterns of Yipu, Suzhou

0 0.5 1m

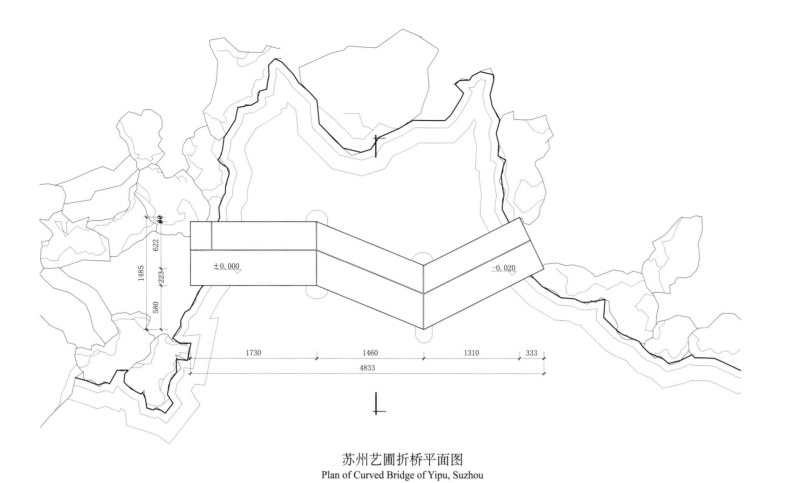

苏州艺圃折桥平面图
Plan of Curved Bridge of Yipu, Suzhou

苏州艺圃折桥立面图
Elevation of Curved Bridge of Yipu, Suzhou

苏州艺圃折桥剖面图
Section of Curved Bridge of Yipu, Suzhou

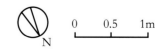

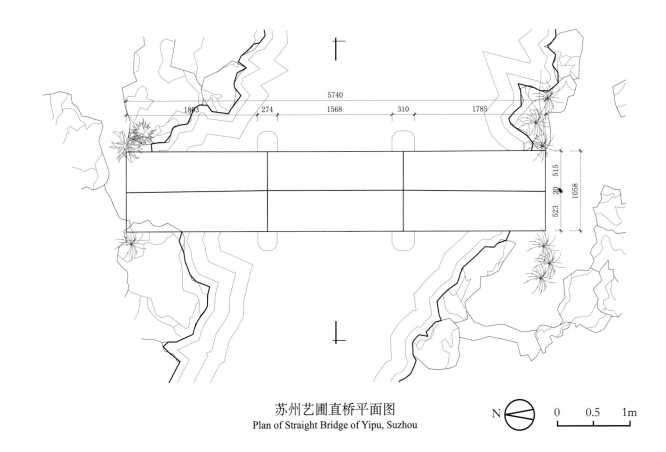

苏州艺圃直桥平面图
Plan of Straight Bridge of Yipu, Suzhou

N 0 0.5 1m

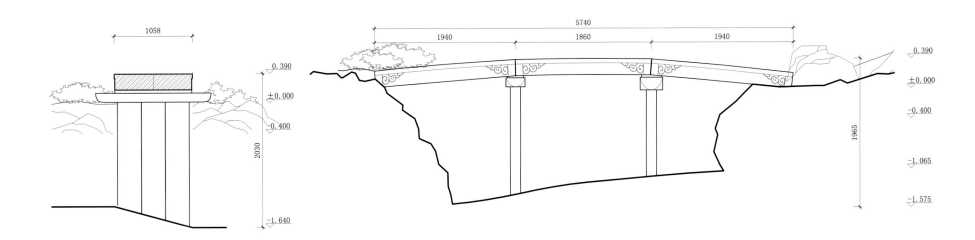

苏州艺圃直桥剖面图
Section of Straight Bridge of Yipu, Suzhou

苏州艺圃直桥立面图
Elevation of Straight Bridge of Yipu, Suzhou

Huanxiu shanzhuang, Suzhou

Sitting at Jingde Road of Suzhou, Huaixiu shanzhuang (Mountain Residence with Embracing Beauty) was built on a site where it was repeatedly repaired and demolished. A garden was first built here by a Wu-yue king named QIAN (907-978). Next, ZHU Changwen built a garden on the same site in Song dynasty. Then in Ming and Qing dynasties, the garden was successively owned by SHEN Shixing, JIANG Ji, BI Yuan, and SUN Shiyi. Finally, it became a shrine of an owner named WANG in 1850 and later renamed Huaixiu shanzhuang.

Covering an area of 2,000 m sq, the garden is known for its rockery hill built by the famous craftsman GE Yuliang in Qing dynasty. The hill is 400 m sq in area, 600 m cubic in volume, and nearing 6 m at its peak. Caves, valleys, crags, cliffs, narrow steps and footbridge define its form. The caves, erected out of rockery arches, appears robust yet organic. The ridges and cliff faces, following a natural contour, are devoid of any traces of handwork. A narrow and meandering pond flows beneath the hill, as if its path was carved out naturally. In fact, every building of the garden is designed, placed and oriented to appreciate this rockery hill. It is out of this layout which derived the phrase "every mountain surface reveals a new form; every step a shifted scene".

苏州环秀山庄

环秀山庄在苏州景德路。地本五代吴越广陵王钱氏（907—978）旧园址，宋时朱长文乐圃也建于此。明清两代先后曾归申时行、蒋楫、毕沅、孙士毅等人。道光末年（1850年）成为汪姓宗祠（义庄）公产，名环秀山庄，俗称汪义庄。

全园面积约2000平方米，以假山著称。山为乾嘉年间叠山名家戈裕良手笔。这是一座30米×30米，占地20米×20米的立体湖石假山，离地面总高不超过6米，而其中有洞壑、涧谷、石壁、悬崖、危径、飞梁等，境界丰富。山洞用拱券原理，以湖石构架，坚实逼真；山峦、石壁则脉理自然，不露人工，浑然天成。山下水池衬托山势陡峭，池窄而水曲，犹如溪流绕山而行，十分灵动。池周的亭廊楼房建造也以欣赏此山为目的，可得到『山形面面看，山景步步移』的效果。

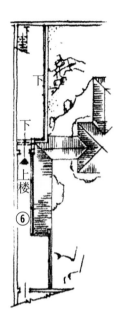
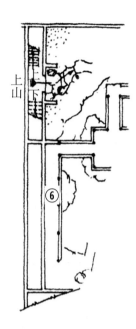

环秀山庄二层平面图
Plan of second floor of Huanxiu shanzhuang, Suzhou

环秀山庄三层平面图
Plan of third floor of Huanxiu shanzhuang, Suzhou

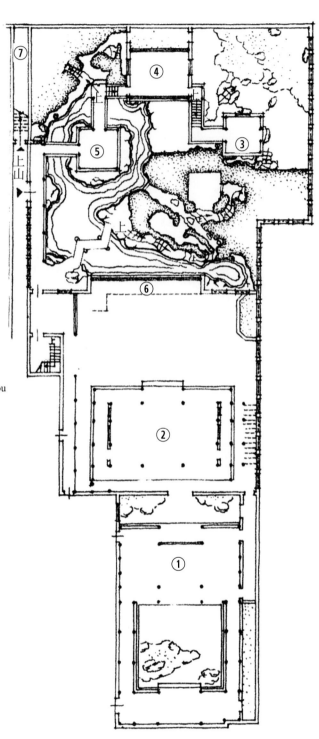

① 有壶堂 Youhutang
② 环秀山庄 Huanxiu shanzhuang
③ 半潭秋水一房山 Bantan qiushui yifangshan
④ 补秋舫 Buqiufang
⑤ 问泉亭（已毁）Wenquanting (Destroyed)
⑥ 廊 Corridor
⑦ 过街楼（已毁）Guojielou (Destroyed)

环秀山庄一层平面图
First floor plan of Huanxiu shanzhuang, Suzhou

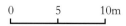

N

0   5   10m

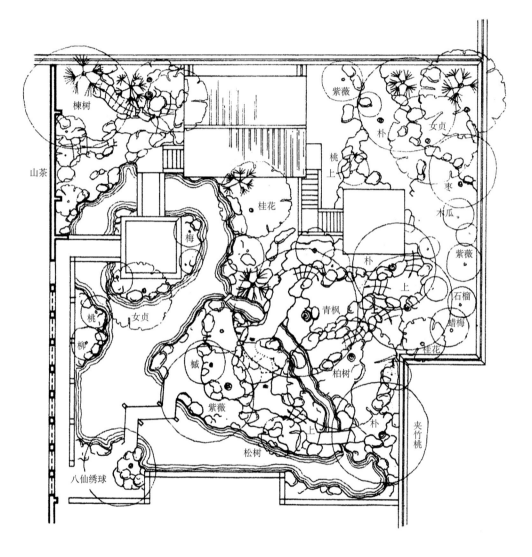

棟树

山茶

桂花

梅

桃

女贞

柳

栎

紫薇

八仙绣球

松树

上

朴

柏树

青枫

朴

上

紫薇

朴

女贞

枣

木瓜

紫薇

石榴

蜡梅

桂花

桃上

夹竹桃

苏州环秀山庄局部屋顶平面图
Partial roof plan of Huanxiu shanzhuang, Suzhou

0　　　5　　　10m

N

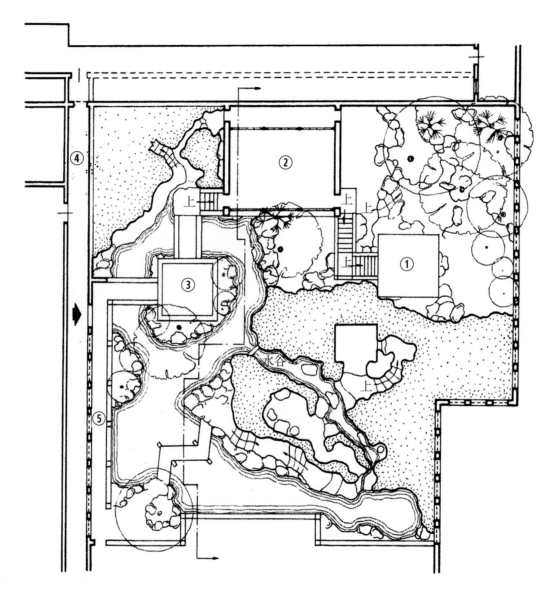

① 半潭秋水一房山 Bantan qiushui yifangshan
② 补秋舫 Buqiufang
③ 问泉亭（已毁）Wenquanting (Destroyed)
④ 过街楼（已毁）Guojielou (Destroyed)
⑤ 廊 Corridor

苏州环秀山庄局部平面图
Part plan of Huanxiu shanzhuang, Suzhou

0　　　　5　　　　10m

N

苏州环秀山庄剖面图
Section of Huanxiu shanzhuang, Suzhou

0 1 2 3m

Ou Garden, Suzhou

Ou Garden (Couple's Garden Retreat), dating to early Qing dynasty, is located at No. 6 Xiao Xinqiao Lane in Suzhou. The garden people see today was rebuilt in late Qing dynasty. The name implies the coexistence of the garden's eastern and western sections.

The smaller, west section is organized as a series of courtyards, defined by libraries and studies, dotted with stacked rockeries, rockery hills, trees and flowers. The larger, east section—covering an area of 2,200 m sq—is centered on a rockery hill and pond, surrounded by buildings, halls, pavilions and galleries. To the section's north stands a two-story building named Chengqu caotang (Secluded Thatched Hall); oriented south, it is the main building of the garden: its width matches the east section's and leads to the residence due west. The hall below serves public functions, the space above for private living. To the section's south are Tinglu (Hearing Splash) Building and Shanshuijian (Mountain and Water) Pavilion. To the east, two pavilions view towards the rockery hill from across the pond. A winding gallery defines the western boundary while connecting buildings across the garden.

The garden's central hill is built using yellow rockery. Expansive in coverage, the rockery's surface and exposed texture were carefully treated across its ridges, cliffs, cracks, valleys and paths. Its top is kept flat without any pavilions in keeping with the hill's scale. However, a building was added later onto the hill, seemingly out of place.

苏州耦园

耦园在苏州小新桥巷6号，清初始建，清末改建成现状。

「耦」与「偶」通，园名意为东西双园并列。

西园较小，实为庭院理景，以书房、书楼周围院落中叠石、叠山、植树、莳花而成。东园较大，占地面积约2200平方米，以山池为中心，环池设楼、堂、亭、榭与游廊。

北面「城曲草堂」是园中主建筑，二层，坐北向南，东西横贯全园；西与住宅相通，楼下厅堂，楼上居室。南面为「听橹楼」与「山水间」水阁，东面隔池望山，设两小亭，西面则以曲廊连通南北两处重要建筑物。园中有黄石假山，将假山的大体大面和石块纹理处理得十分妥当，峰峦、峭壁、裂隙、峡谷、磴道，各尽其趣。山顶只做平台而不建亭，山上石室为后世所加，似与全山风格不符。无琐碎堆砌之弊。

① 入口 Entrance
② 门厅 Entrance Hall
③ 轿厅 Sedan Chair Hall
④ 载酒堂 Zaijiutang
⑤ 楼厅 Lou Hall
⑥ 纫兰室 Renlanshi
⑦ 织帘老屋 Zhilian laowu
⑧ 藏书楼 Library Building
⑨ 鹤寿亭 Heshou Pavilion
⑩ 无俗韵轩 Wusu yunxuan
⑪ 山水间 Shanshuijian (Mountain and Water) Pavilion
⑫ 藤花舫 Tenghuafang
⑬ 城曲草堂 Chengqu caotang (Secluded Thatched Hall)
⑭ 储香馆 Chuxiangguan
⑮ 还砚斋 Huanyanzhai
⑯ 吾爱亭 Wuaiting
⑰ 听橹楼 Tinglu (Hearing Splash) Building
⑱ 魁星阁 Kuixingge
⑲ 樨廊 Xilang
⑳ 望月亭 Wangyueting
㉑ 天井 Small Yard
㉒ 游船码头 Pier
㉓ 后附房 Attached room
㉔ 筠廊 Yunlang
㉕ 售票处 Ticket office

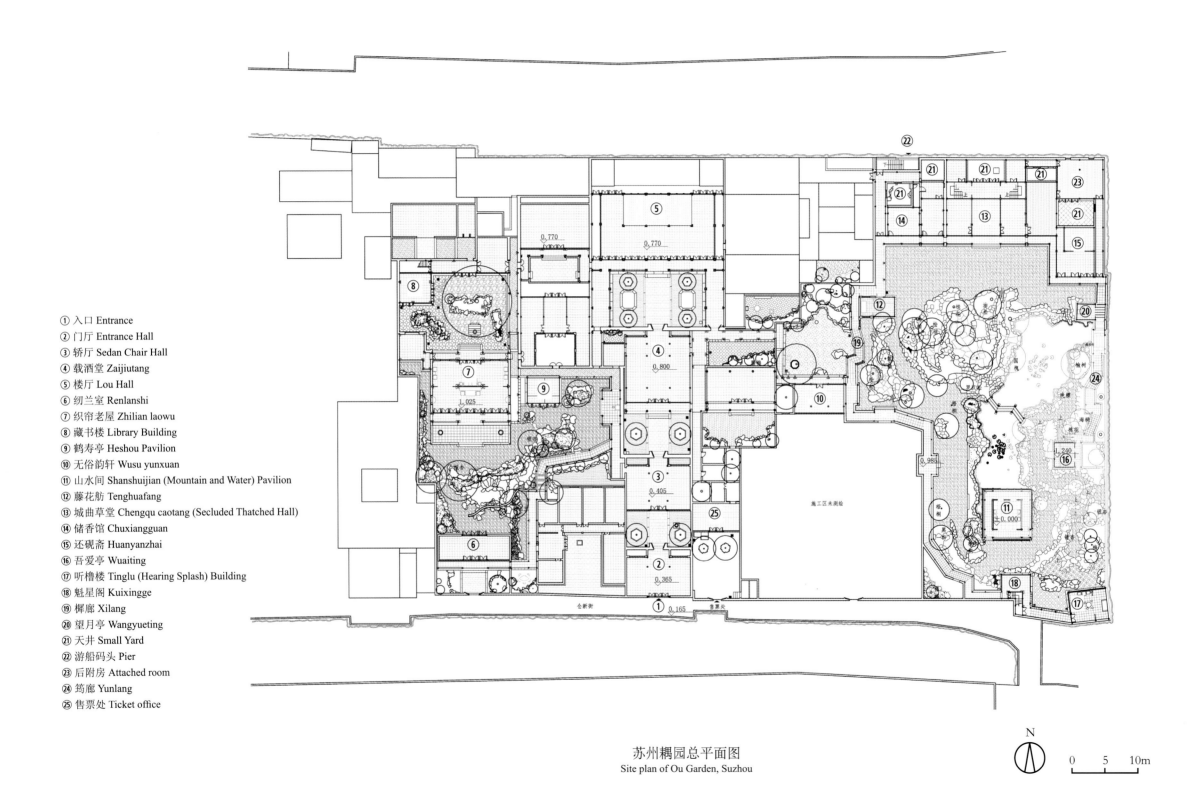

苏州耦园总平面图
Site plan of Ou Garden, Suzhou

N

0   5   10m

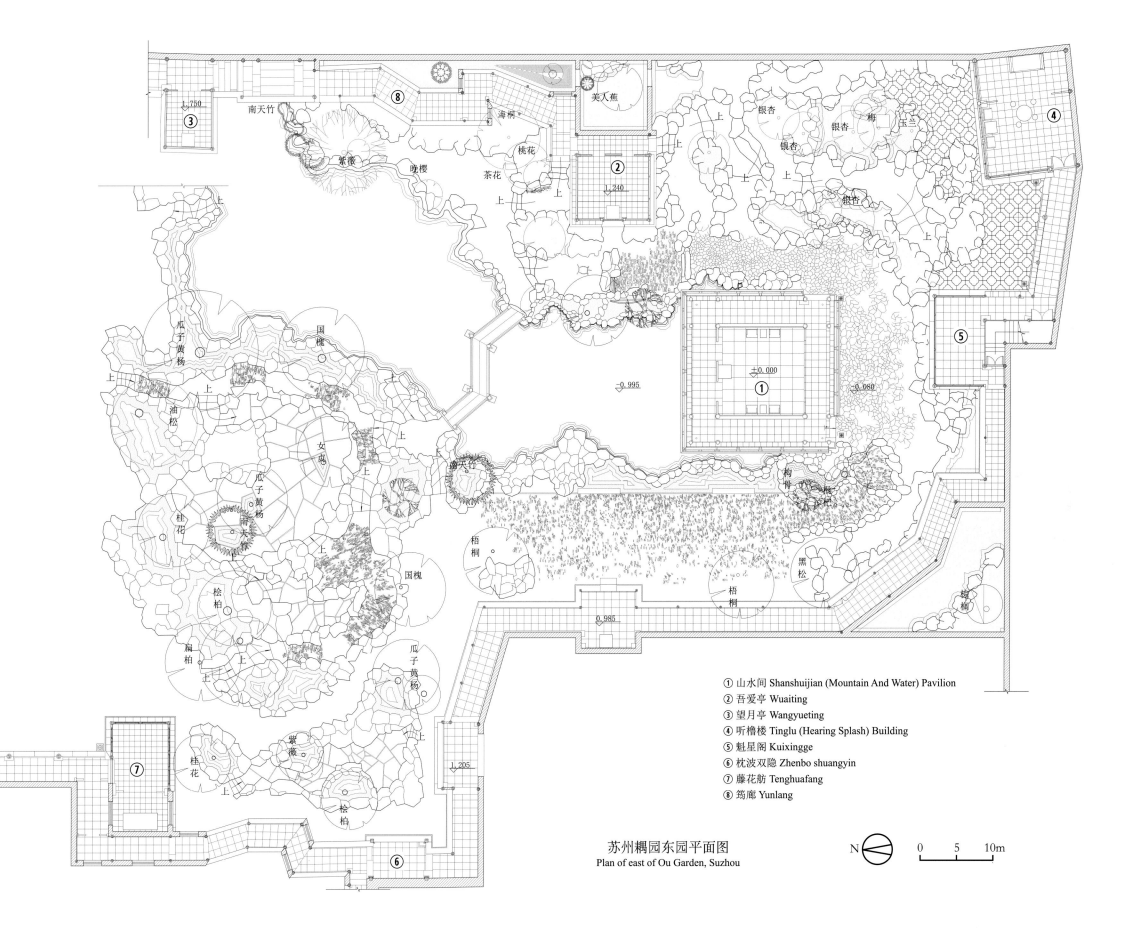

1.750
③
南天竹
海桐
⑧
美人蕉
④
紫薇
晚樱
桃花
茶花
银杏
银杏
银杏
梅
玉兰
②
1.240
上
上
上
银杏
瓜子黄杨
国槐
±0.000
①
⑤
油松
-0.995
-0.080
女贞
上
南天竹
瓜子黄杨
南天竹
构橘
桂花
枇杷
梧桐
黑松
瓜子黄杨
国槐
梧桐
桧柏
檀桐
扁柏
瓜子黄杨
0.985
桂花
⑦
紫薇
1.205
桧柏
⑥

① 山水间 Shanshuijian (Mountain And Water) Pavilion
② 吾爱亭 Wuaiting
③ 望月亭 Wangyueting
④ 听橹楼 Tinglu (Hearing Splash) Building
⑤ 魁星阁 Kuixingge
⑥ 枕波双隐 Zhenbo shuangyin
⑦ 藤花舫 Tenghuafang
⑧ 筠廊 Yunlang

苏州耦园东园平面图
Plan of east of Ou Garden, Suzhou

N
0 5 10m

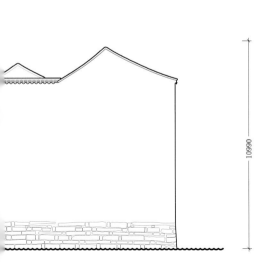

游船码头

10990

10630

仓新街　　主（南）入口　售票处　　　工作入口

苏州耦园屋顶平面图
Roof plan of Ou Garden, Suzhou

N

0　　5　　10m

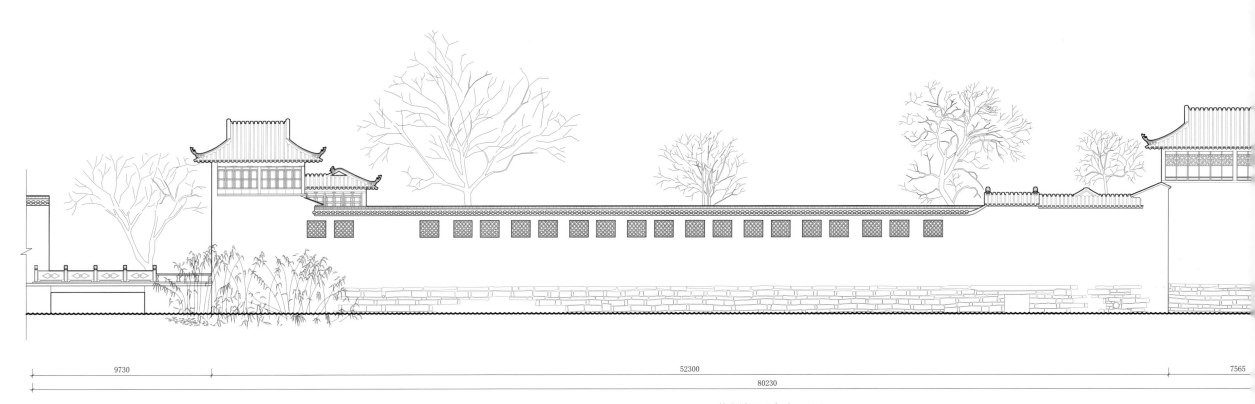

2575 1810 3945 12955 1350 4170 2195 7560

2574 1808 5343 4008 4120 3430 1435 4079 3674 4760 1325

苏州耦园东花园剖面图

Section of east of Ou Garden, Suzhou

0 5 10m

9730 7565 52300 7565

80230

0 5 10m

苏州耦园东立面图

East elevation of Ou Garden, Suzhou

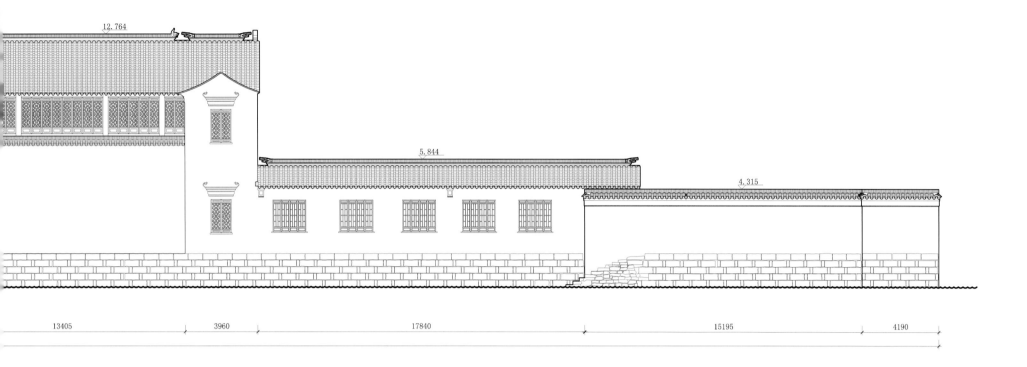

12.764

5.844

4.315

| 13405 | 3960 | 17840 | 15195 | 4190 |

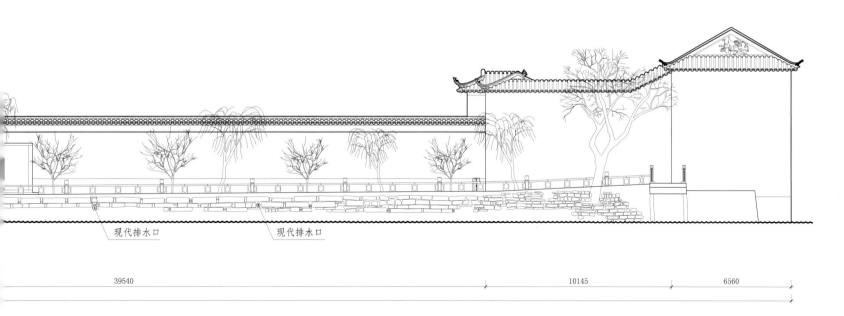

现代排水口　　　　现代排水口

| 39540 | 10145 | 6560 |

0　　5　　10m

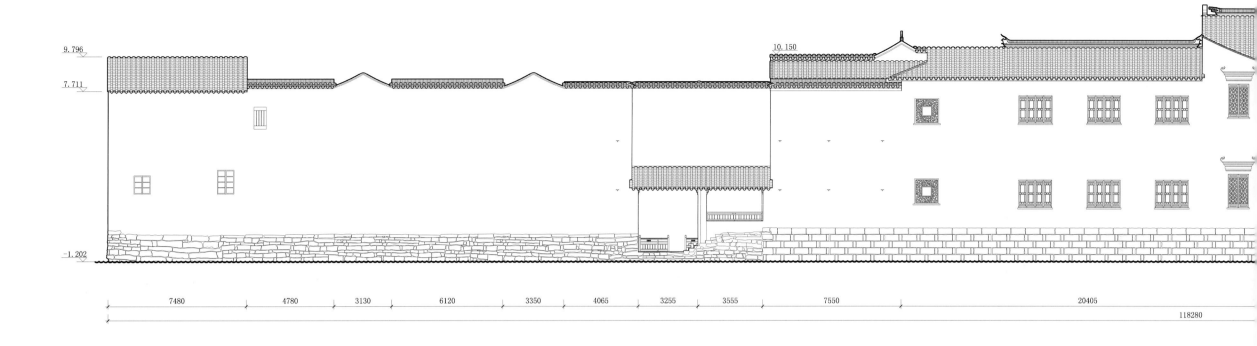

9.796

7.711

9.150

10.150

-1.202

| 7480 | 4780 | 3130 | 6120 | 3350 | 4065 | 3255 | 3555 | 7550 | 20405 |

118280

苏州耦园北立面图
North elevation of Ou Garden, Suzhou

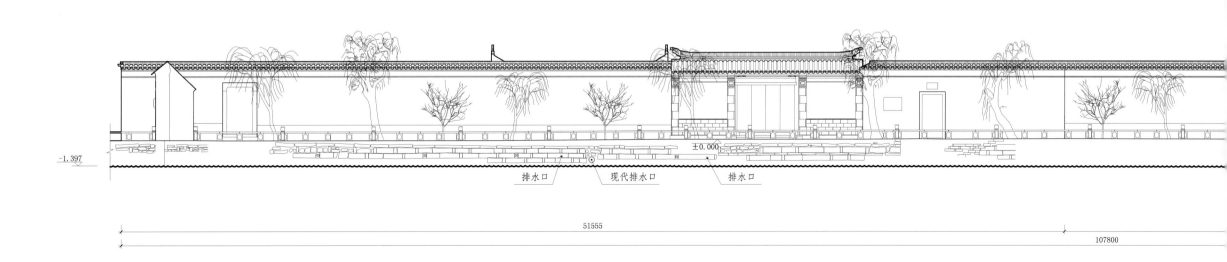

-1.397

±0.000

排水口　　现代排水口　　排水口

51555

107800

苏州耦园南立面图
South elevation of Ou Garden, Suzhou

苏州耦园黄石假山南立面图
South elevation of yellow rockery hill of Ou Garden, Suzhou

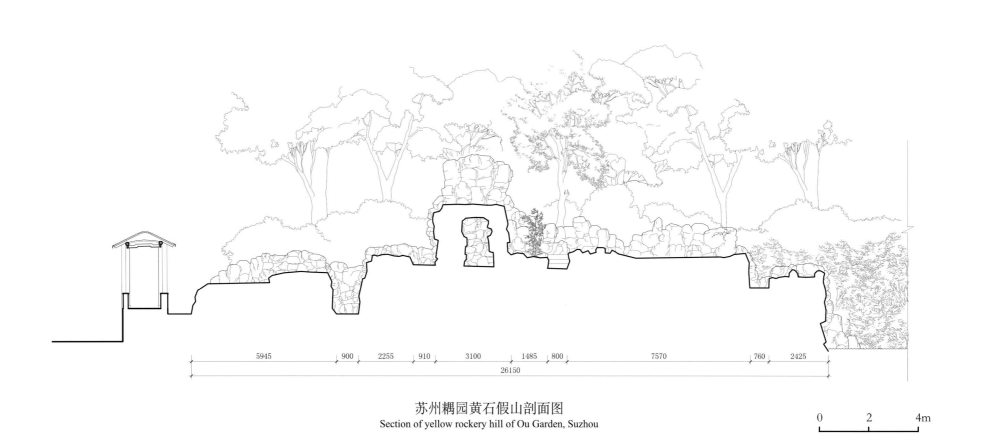

苏州耦园黄石假山剖面图
Section of yellow rockery hill of Ou Garden, Suzhou

0　　2　　4m

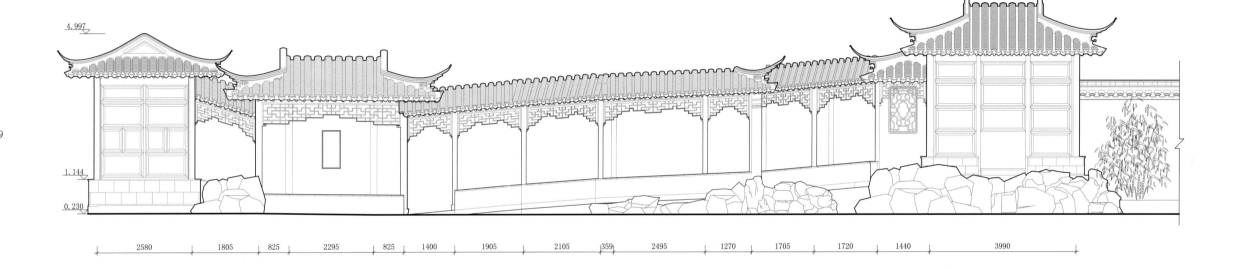

苏州耦园筠廊立面图
Elevation of Yunlang of Ou Garden, Suzhou

0　1　2m

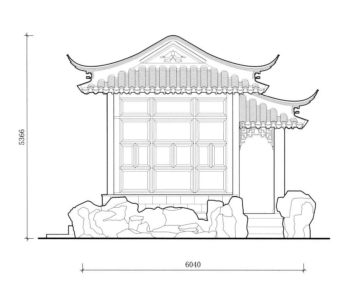

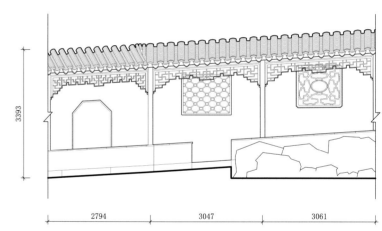

苏州耦园望月亭剖面图
Section of Wangyueting of Ou Garden,Suzhou

苏州耦园望月亭立面图
Elevation of Wangyueting of Ou Garden,Suzhou

苏州耦园樨廊立面图（由南至北 1 段）
Elevation of Xilang (first section from south to north) of Ou Garden,Suzhou

0　　1　　2m

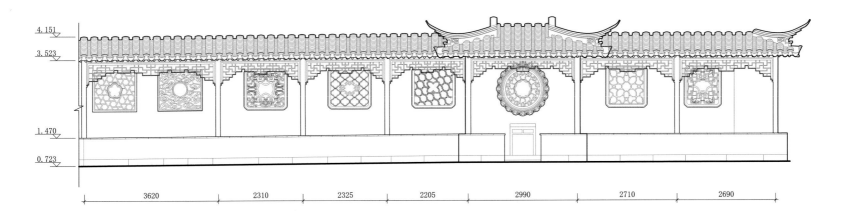

4.151
3.523
1.470
0.723

3620　2310　2325　2205　2990　2710　2690

苏州耦园樨廊立面图（由南至北 2 段）

Elevation of Xilang (second section from south to north) of Ou Garden, Suzhou

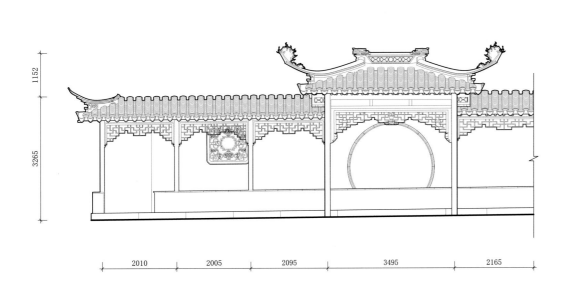

1152
3265

2010　2005　2095　3495　2165

苏州耦园樨廊立面图（由南至北 3 段）

Elevation of Xilang (third section from south to north) of Ou Garden, Suzhou

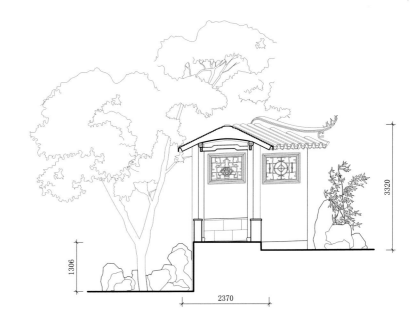

3320
1306

2370

苏州耦园樨廊剖面图

Section of Xilang of Ou Garden, Suzhou

0　1　2m

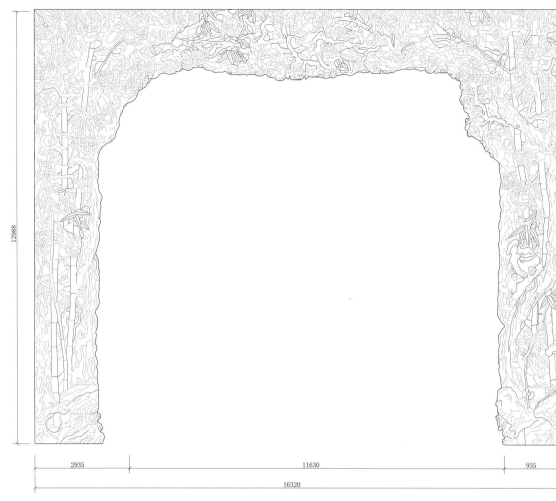

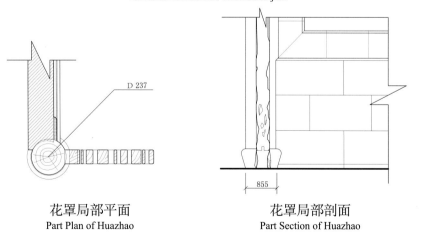

山水间花罩立面
Elevation of Huazhao of Shanshuijian

花罩局部平面
Part Plan of Huazhao

花罩局部剖面
Part Section of Huazhao

苏州耦园山水间花罩详图
Detail of Huazhao of Shanshuijian of Ou Garden, Suzhou

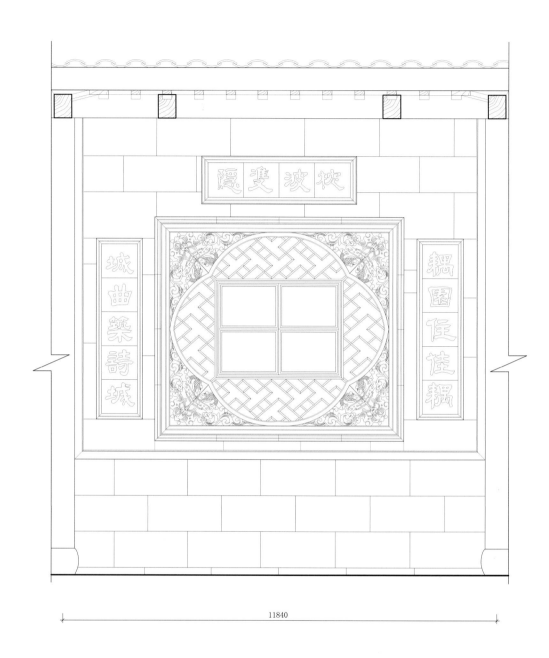

苏州耦园枕波双隐大样图
Wall inscription of Zhenbo shuangyin Pavilion of Ou Garden, Suzhou

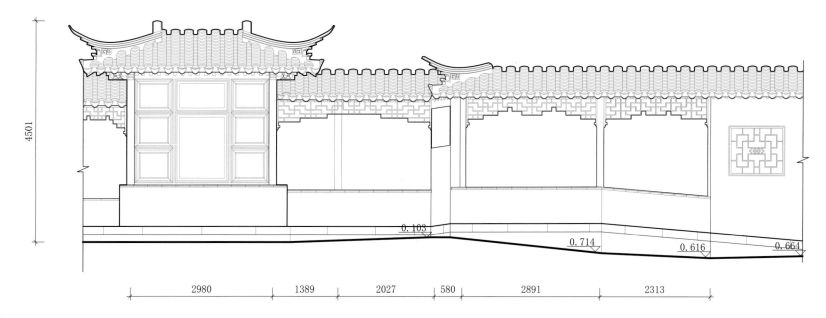

4501

0.103

0.714   0.616   0.664

2980 | 1389 | 2027 | 580 | 2891 | 2313

苏州耦园樨廊立面图（由南至北 4 段）

Elevation of Xilang (fourth section from south to north) of Ou Garden, Suzhou

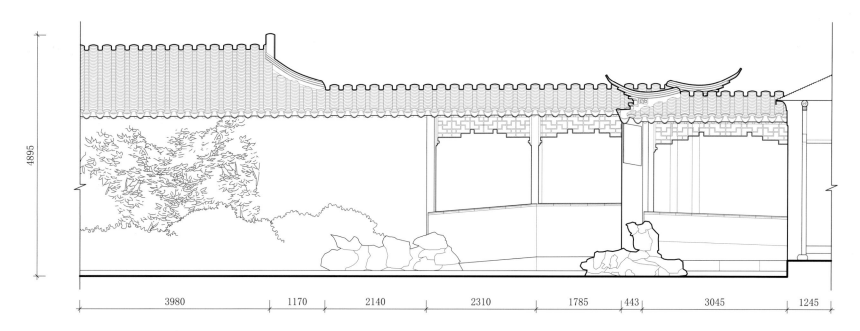

4895

3980 | 1170 | 2140 | 2310 | 1785 | 443 | 3045 | 1245

苏州耦园樨廊立面图（由南至北 5 段）

Elevation of Xilang (fifth section from south to north) of Ou Garden, Suzhou

0   1   2m

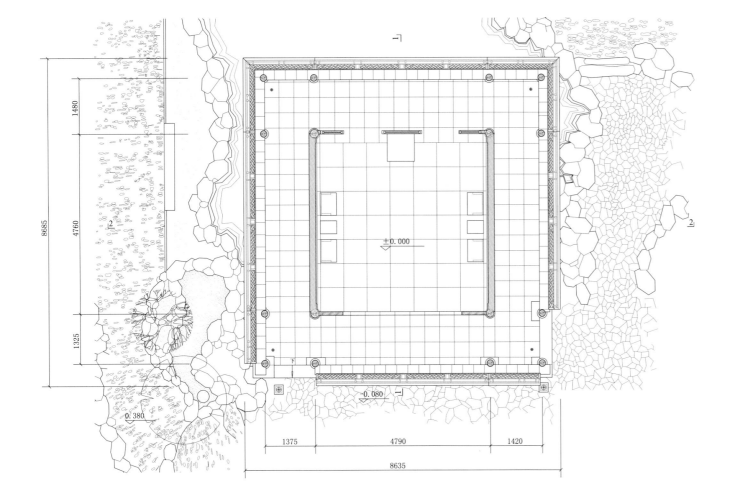

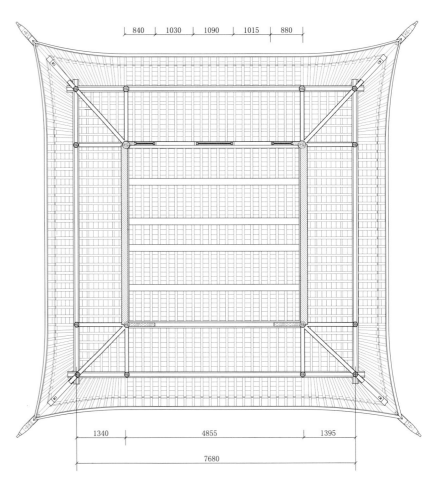

苏州耦园山水间平面图
Site plan of Shanshuijian of Ou Garden, Suzhou

0 1 2m

N

苏州耦园山水间仰视平面图
Reflected ceiling plan of Shanshuijian of Ou Garden, Suzhou

768
510
696
2470
1045
455

−0.080

−0.935

8085

608
845
390
696
2470
1045
455

7680

苏州耦园山水间 1−1 剖面图
Section 1-1 of Shanshuijian of Ou Garden, Suzhou

0　　1　　2m

苏州耦园山水间 2−2 剖面图
Section 2-2 of Shanshuijian of Ou Garden, Suzhou

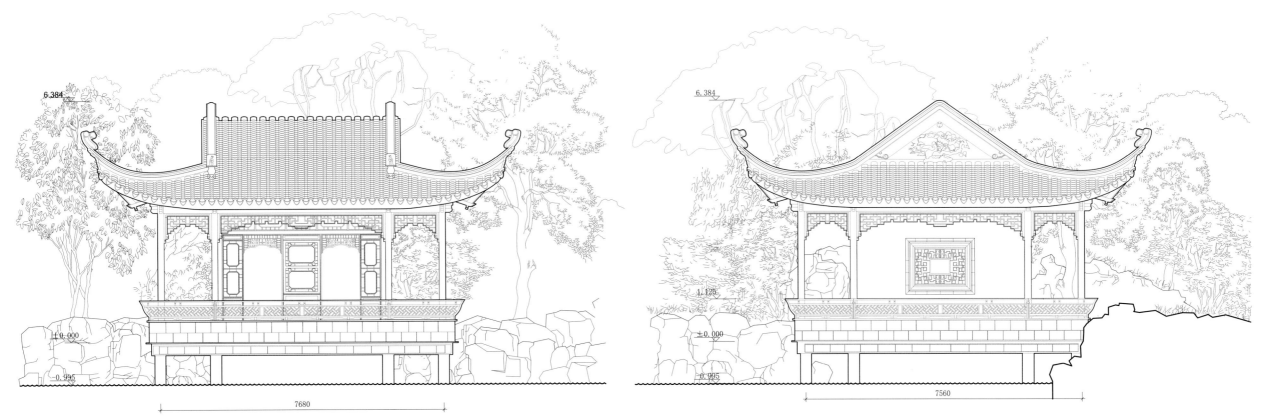

6.384

±0.000

-0.995

7680

苏州耦园山水间北立面图
North elevation of Shanshuijian of Ou Garden, Suzhou

0　　1　　2m

6.384

1.125

±0.000

-0.995

7560

苏州耦园山水间西立面图
West elevation of Shanshuijian of Ou Garden, Suzhou

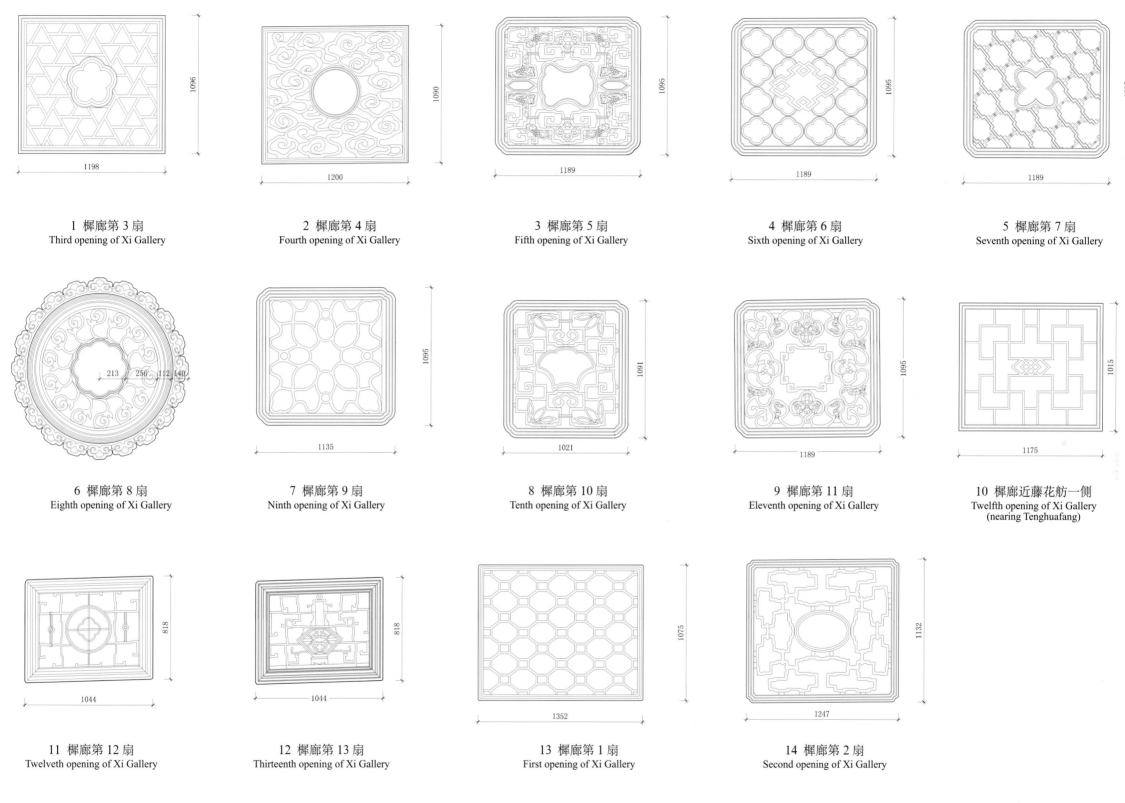

1 樨廊第 3 扇
Third opening of Xi Gallery

2 樨廊第 4 扇
Fourth opening of Xi Gallery

3 樨廊第 5 扇
Fifth opening of Xi Gallery

4 樨廊第 6 扇
Sixth opening of Xi Gallery

5 樨廊第 7 扇
Seventh opening of Xi Gallery

6 樨廊第 8 扇
Eighth opening of Xi Gallery

7 樨廊第 9 扇
Ninth opening of Xi Gallery

8 樨廊第 10 扇
Tenth opening of Xi Gallery

9 樨廊第 11 扇
Eleventh opening of Xi Gallery

10 樨廊近藤花舫一侧
Twelfth opening of Xi Gallery
(nearing Tenghuafang)

11 樨廊第 12 扇
Twelveth opening of Xi Gallery

12 樨廊第 13 扇
Thirteenth opening of Xi Gallery

13 樨廊第 1 扇
First opening of Xi Gallery

14 樨廊第 2 扇
Second opening of Xi Gallery

苏州耦园花窗大样图（以听橹楼为起点）
Perforated opening patterns (number starting from Tinglulou) of Ou Garden, Suzhou

0　　　0.5　　　1m

① 魁星阁 Kuixingge
② 听橹楼 Tinglu (Hearing Splash) Building

苏州耦园魁星阁和听橹楼平面图
Site plan of Kuixingge and Tinglu Building of Ou Garden, Suzhou

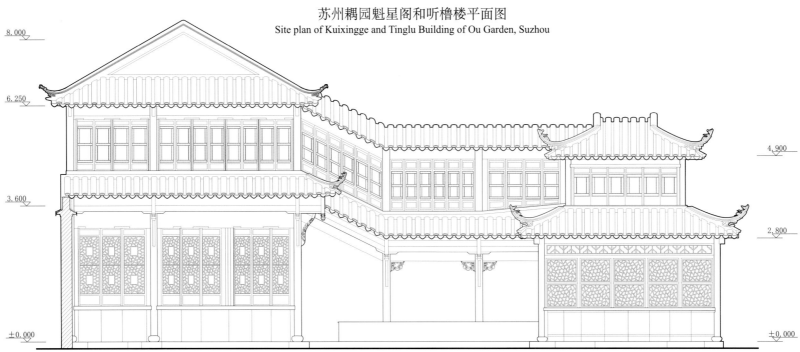

苏州耦园魁星阁和听橹楼北立面图
North elevation of Kuixingge and Tinglu Building of Ou Garden, Suzhou

苏州耦园魁星阁和听橹楼 1—1 剖面图
Section 1-1 of Kuixingge and Tinglu Building of Ou Garden, Suzhou

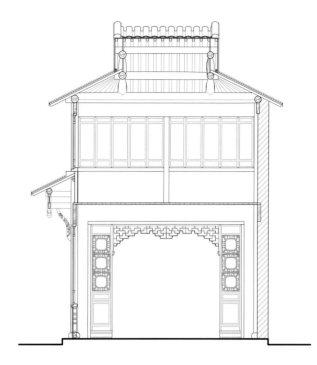

苏州耦园魁星阁和听橹楼 2—2 剖面图
Section 2-2 of Kuixingge and Tinglu Building of Ou Garden, Suzhou

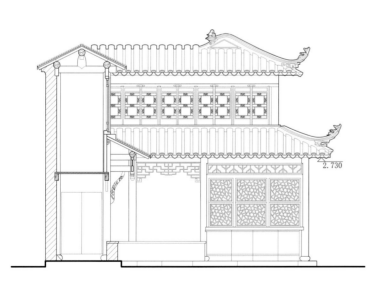

苏州耦园魁星阁和听橹楼 3—3 剖面图
Section 3-3 of Kuixingge and Tinglu Building of Ou Garden, Suzhou

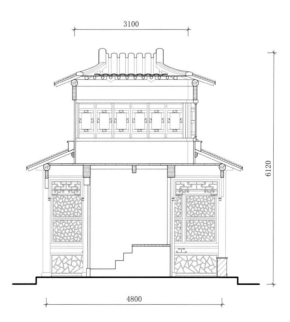

苏州耦园魁星阁和听橹楼 4—4 剖面图
Section 4-4 of Kuixingge and Tinglu Building of Ou Garden, Suzhou

5950

6120

6340

苏州耦园魁星阁和听橹楼 5-5 剖面图
Section 5-5 of Kuixingge and Tinglu Building of Ou Garden, Suzhou

0   1   2m

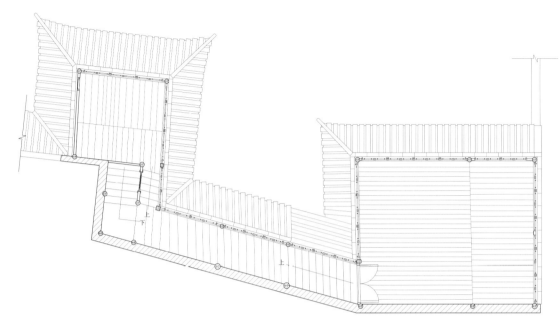

苏州耦园魁星阁和听橹楼二层平面图
Plan of second floor of Kuixingge and Tinglu Building of Ou Garden, Suzhou

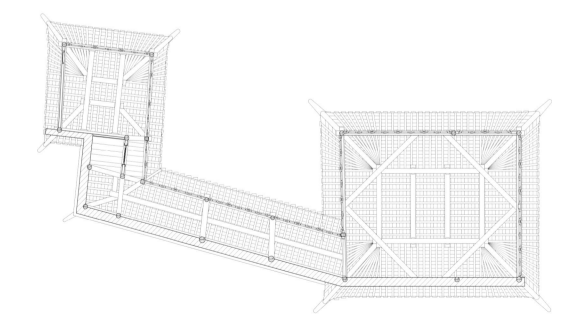

苏州耦园魁星阁和听橹楼仰视平面图
Reflected ceiling plan of Kuixingge and Tinglu Building of Ou Garden, Suzhou

0   1   2m

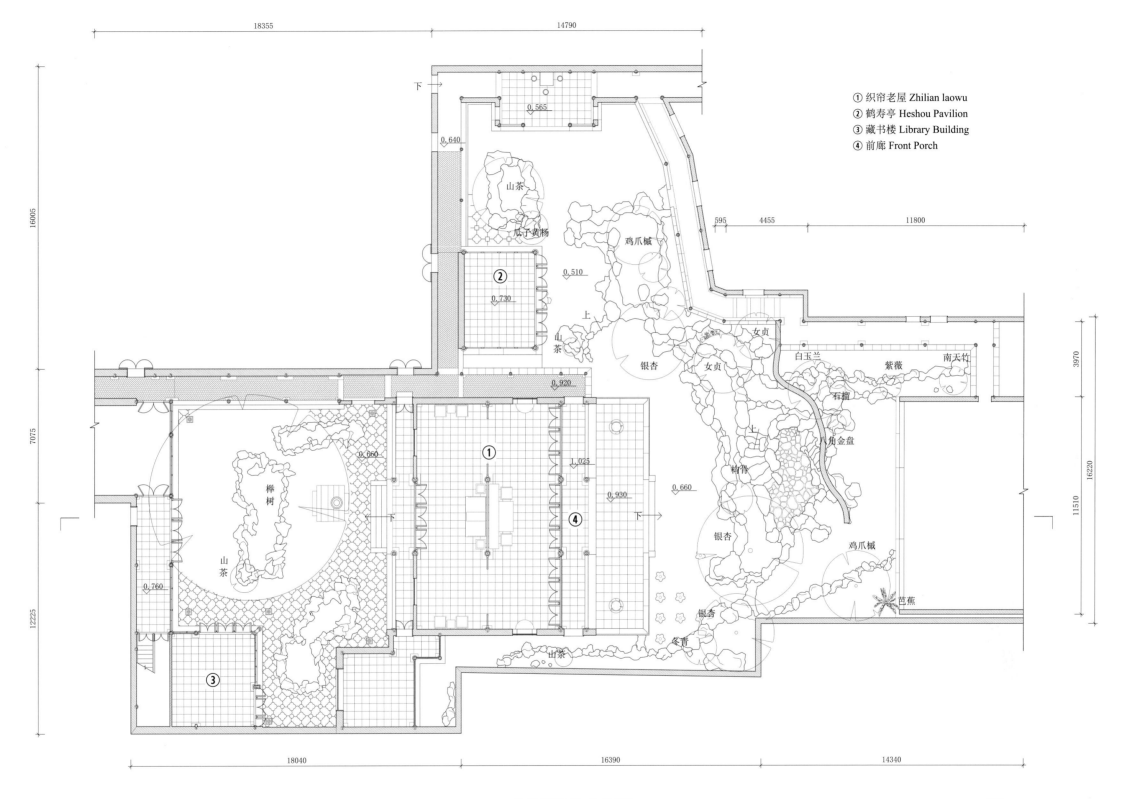

① 织帘老屋 Zhilian laowu
② 鹤寿亭 Heshou Pavilion
③ 藏书楼 Library Building
④ 前廊 Front Porch

山茶

瓜子黄杨

鸡爪槭

②

山茶

银杏

女贞

女贞

白玉兰

紫薇

南天竹

石榴

八角金盘

①

榉树

枸骨

④

银杏

山茶

鸡爪槭

③

银杏

山茶

冬青

芭蕉

苏州耦园西园平面图
Plan of west of Ou Garden, Suzhou

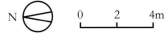

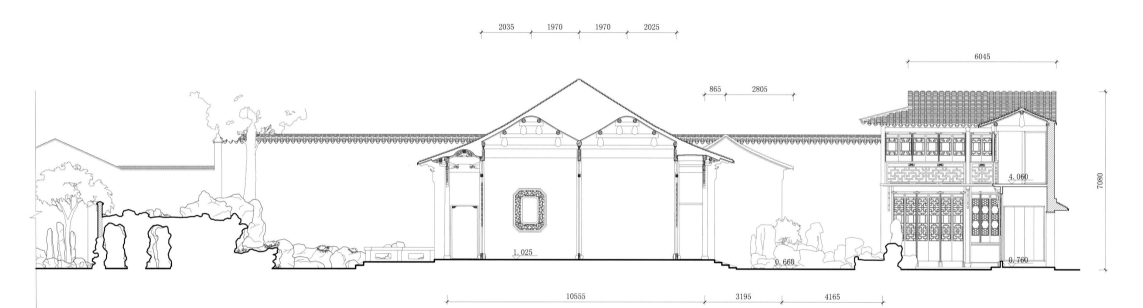

2035　1970　1970　2025

6045

865　2805

4.060

7080

1.025

0.660　　0.760

10555　　　3195　　4165

苏州耦园西园剖面图
Section of west of Ou Garden, Suzhou

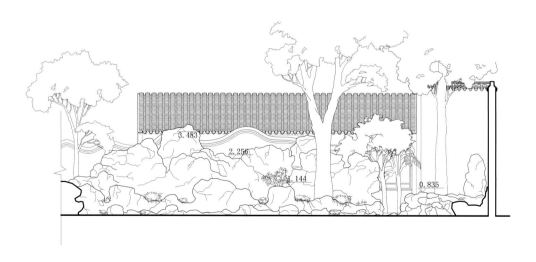

3.483

2.256

1.144

0.835

苏州耦园西园假山北立面图
North elevation of rockery hill of west of Ou Garden, Suzhou

0　　2　　4m

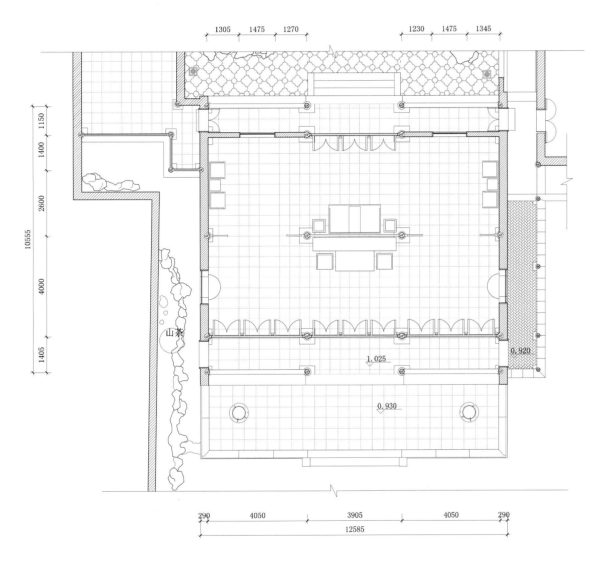

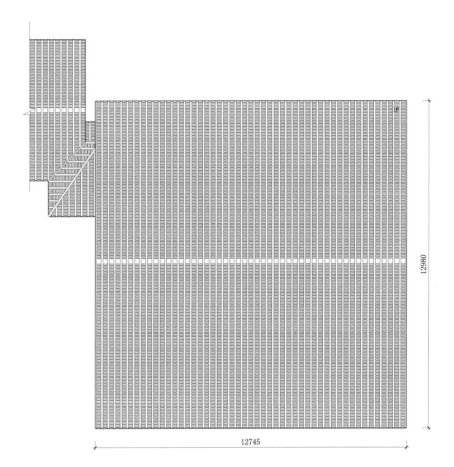

苏州耦园织帘老屋平面图
Site plan of Zhilian laowu of Ou Garden, Suzhou

苏州耦园织帘老屋屋顶平面图
Roof plan of Zhilian laowu of Ou Garden, Suzhou

0 2 4m

N

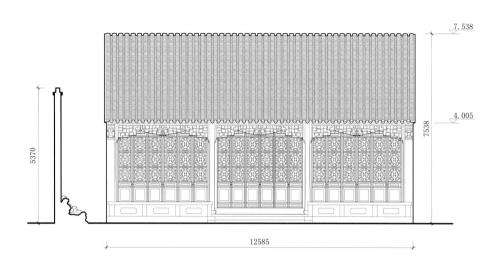

苏州耦园织帘老屋南立面图
South elevation of Zhilian laowu of Ou Garden, Suzhou

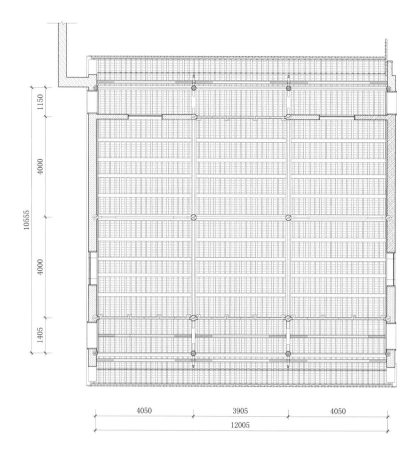

苏州耦园织帘老屋仰视平面图
Reflected ceiling plan of Zhilian laowu of Ou Garden, Suzhou

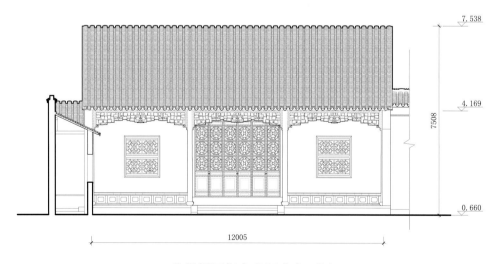

苏州耦园织帘老屋北立面图
North elevation of Zhilian laowu of Ou Garden, Suzhou

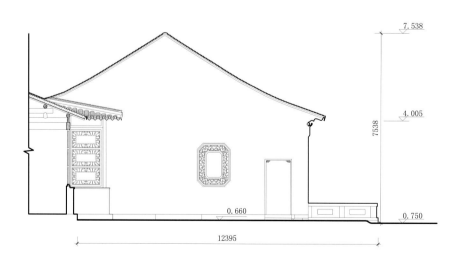

苏州耦园织帘老屋西立面图
West elevation of Zhilian laowu of Ou Garden, Suzhou

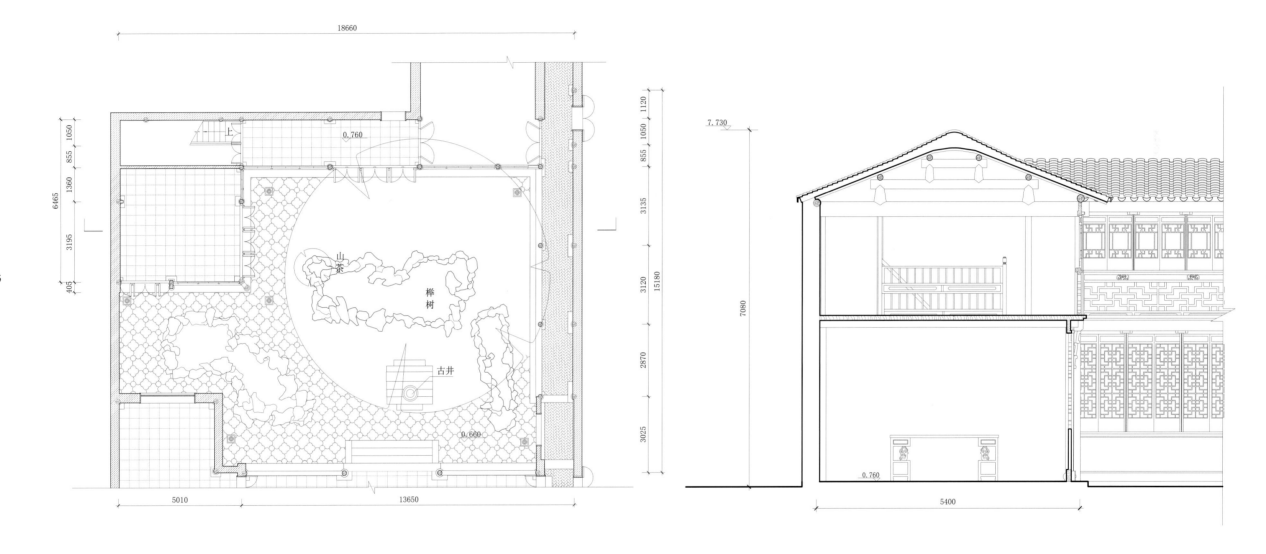

苏州耦园藏书楼一层平面图
Plan of first floor of Library Building of Ou Garden, Suzhou

N

苏州耦园藏书楼剖面图
Section of Library Building of Ou Garden, Suzhou

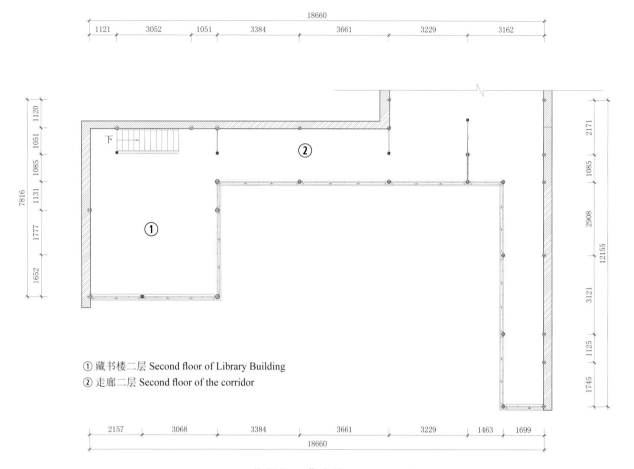

① 藏书楼二层 Second floor of Library Building
② 走廊二层 Second floor of the corridor

苏州耦园藏书楼二层平面图
Plan of second floor of Library Building of Ou Garden, Suzhou

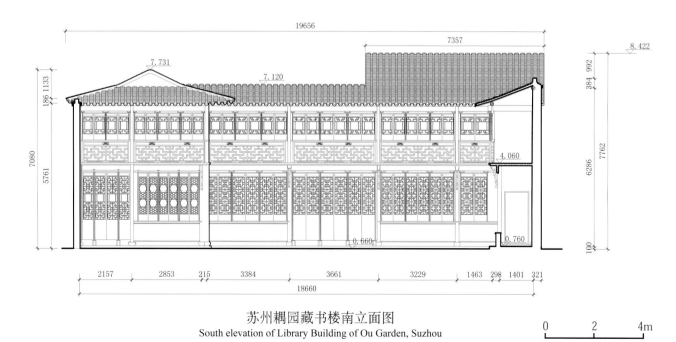

苏州耦园藏书楼南立面图
South elevation of Library Building of Ou Garden, Suzhou

0   2   4m

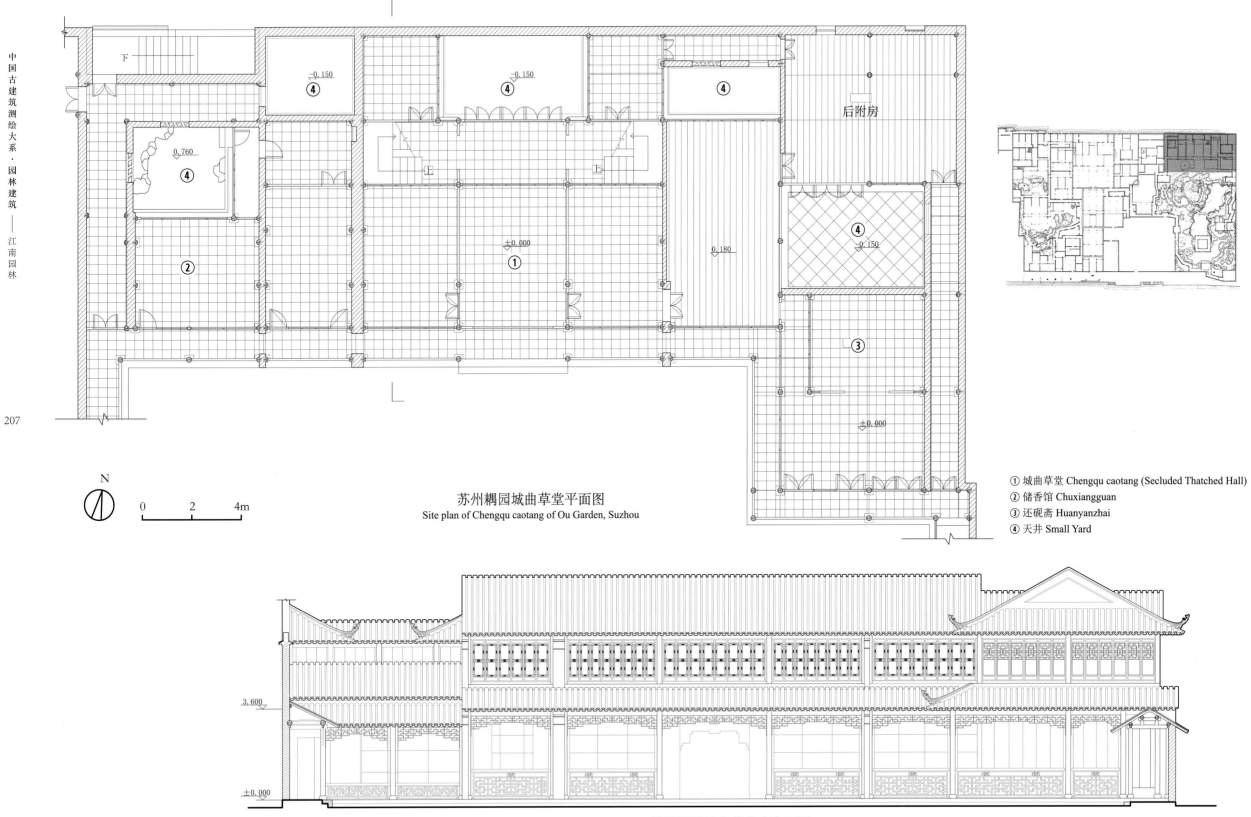

后附房

N

0　2　4m

苏州耦园城曲草堂平面图
Site plan of Chengqu caotang of Ou Garden, Suzhou

① 城曲草堂 Chengqu caotang (Secluded Thatched Hall)
② 储香馆 Chuxiangguan
③ 还砚斋 Huanyanzhai
④ 天井 Small Yard

苏州耦园城曲草堂南立面图
South elevation of Chengqu caotang of Ou Garden, Suzhou

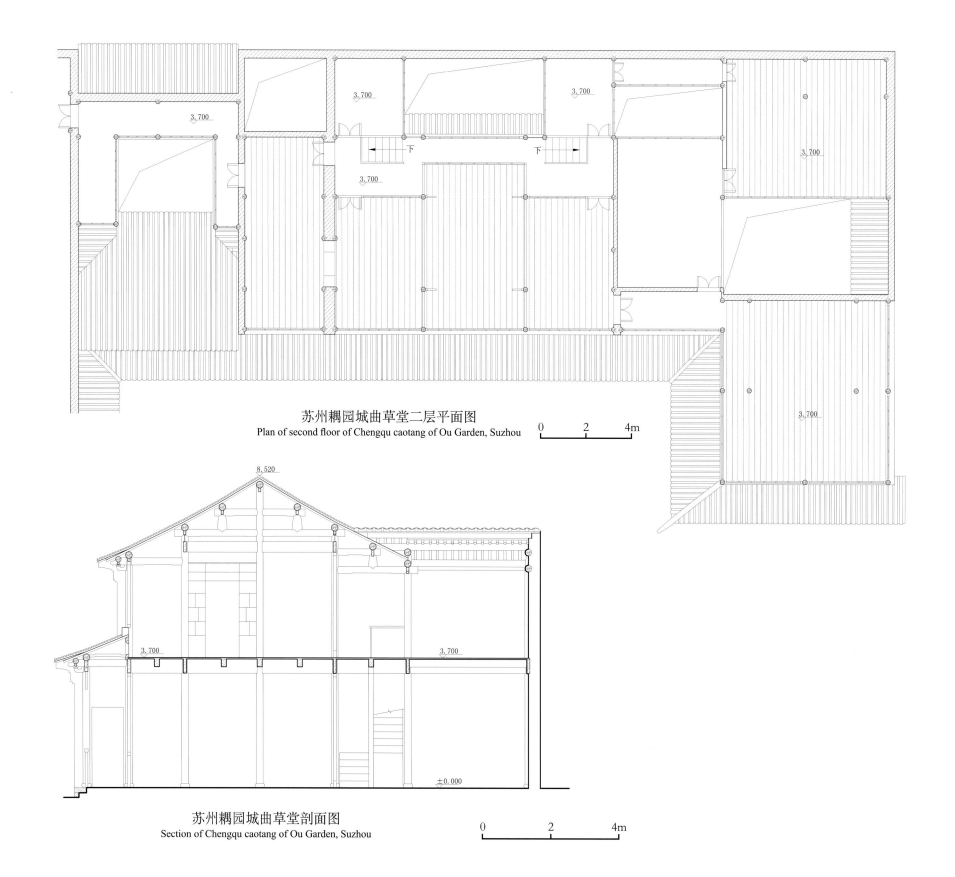

苏州耦园城曲草堂二层平面图
Plan of second floor of Chengqu caotang of Ou Garden, Suzhou

0   2   4m

苏州耦园城曲草堂剖面图
Section of Chengqu caotang of Ou Garden, Suzhou

0   2   4m

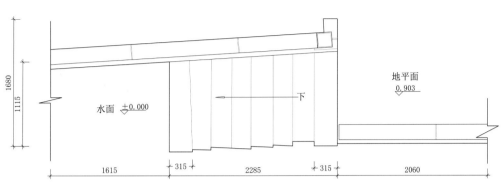

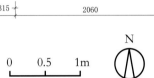

苏州耦园南码头平面图
Plan of South pier of Ou Garden, Suzhou

0　0.5　1m

N

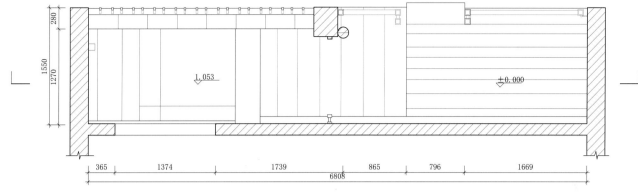

苏州耦园北码头平面图
Plan of North pier of Ou Garden, Suzhou

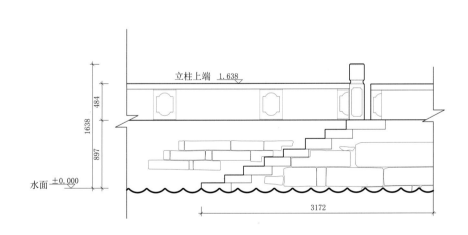

苏州耦园南码头立面图
Elevation of South pier of Ou Garden, Suzhou

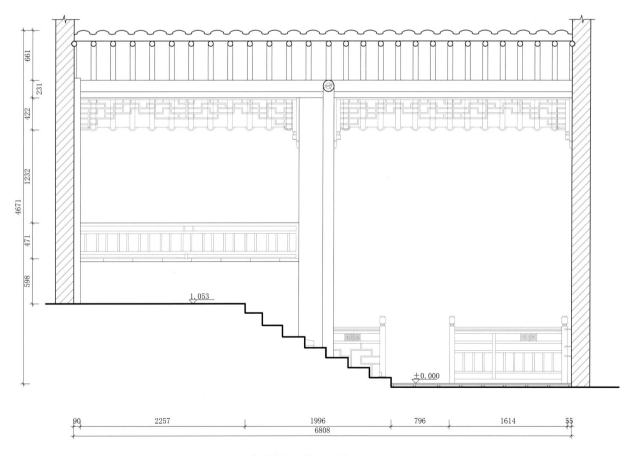

苏州耦园北码头剖面图
Section of North pier of Ou Garden, Suzhou

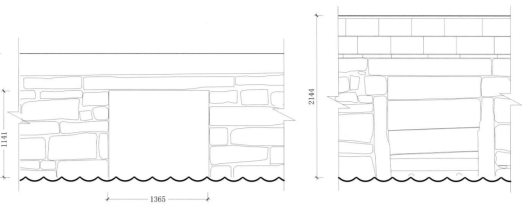

苏州耦园涵洞外立面图
Outside elevation of Culvert of Ou Garden, Suzhou

苏州耦园涵洞内立面图
Inside elevation of Culvert of Ou Garden, Suzhou

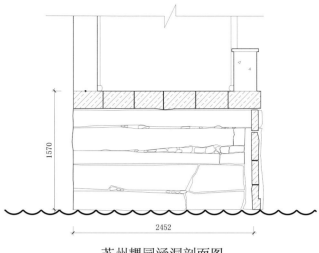

苏州耦园涵洞剖面图
Section of Culvert of Ou Garden, Suzhou

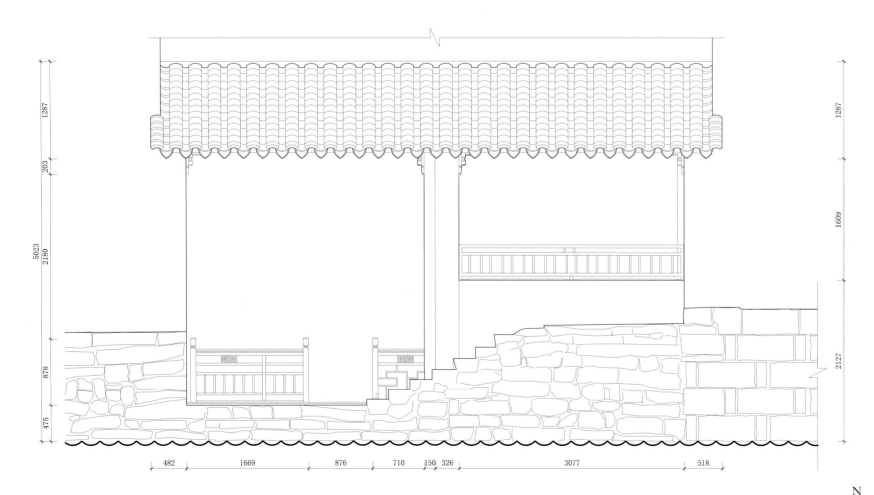

苏州耦园北码头立面图
Elevation of North pier of Ou Garden, Suzhou

N

0   0.5   1m

苏州耦园涵洞平面图
Plan of Culvert of Ou Garden, Suzhou

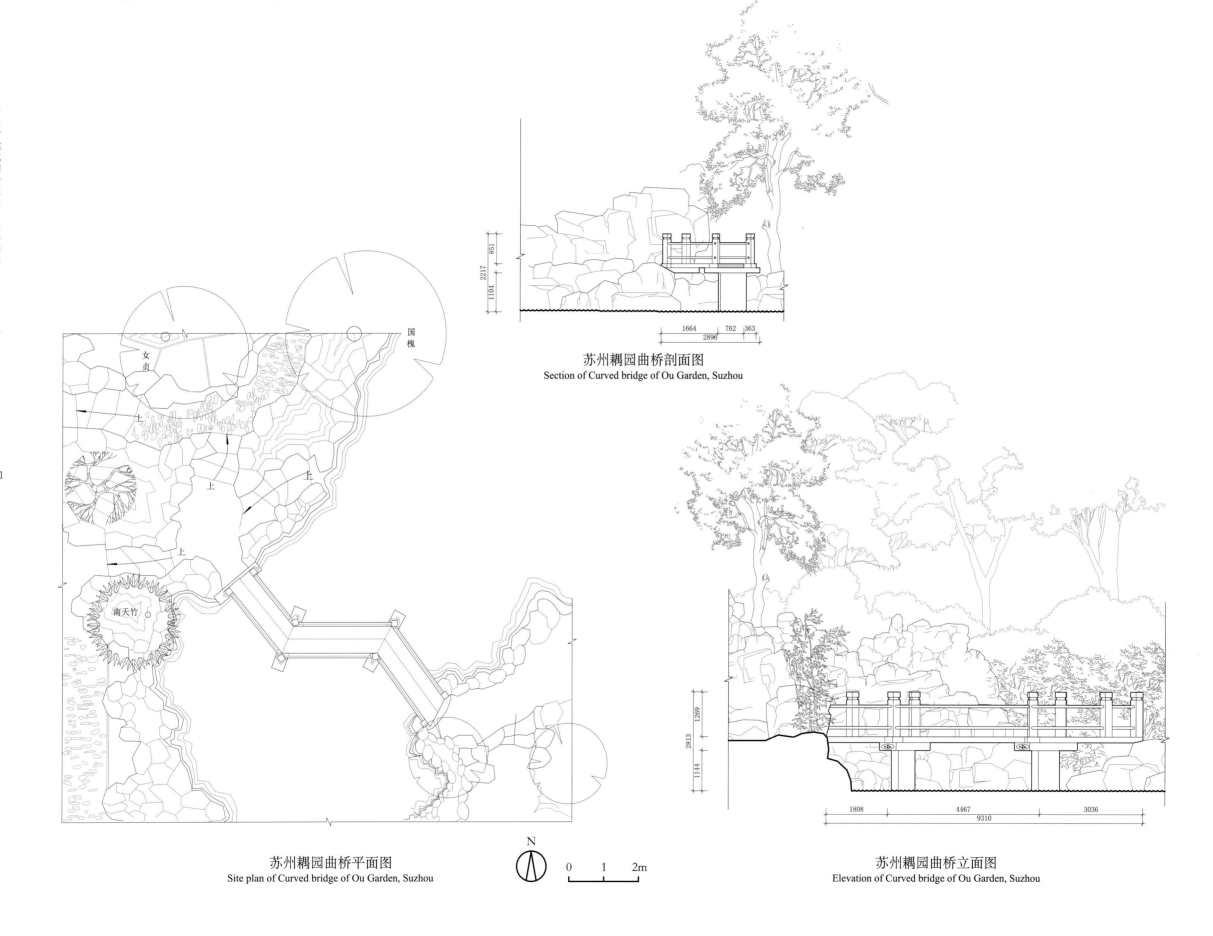

国槐

女贞

南天竹

上

上

上

苏州耦园曲桥剖面图
Section of Curved bridge of Ou Garden, Suzhou

2217
851
1104

1664　762　363
2896

苏州耦园曲桥平面图
Site plan of Curved bridge of Ou Garden, Suzhou

N

0　1　2m

苏州耦园曲桥立面图
Elevation of Curved bridge of Ou Garden, Suzhou

2813
1269
1144

1808　4467　3036
9310

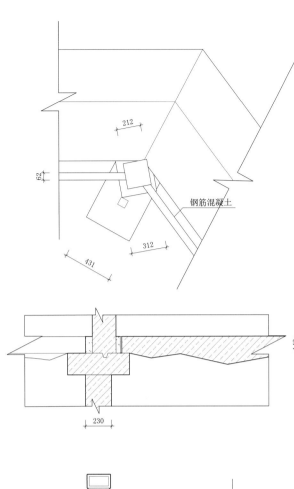

钢筋混凝土

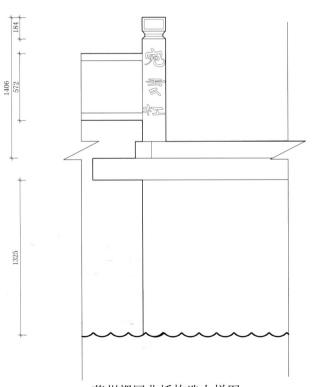

苏州耦园曲桥构造大样图
Construction of Curved bridge of Ou Garden, Suzhou

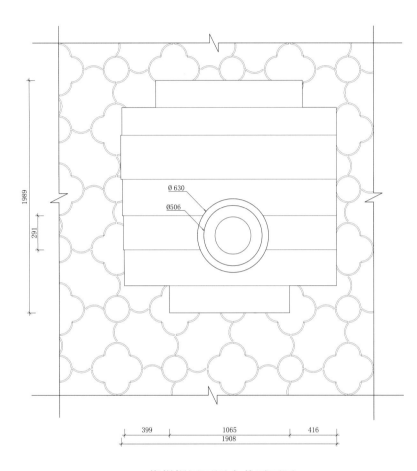

Ø630
Ø506

苏州耦园西园水井平面图
Plan of Well of west of Ou Garden, Suzhou

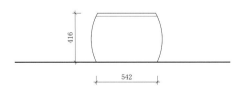

苏州耦园西园水井立面图
Elevation of Well of west of Ou Garden, Suzhou

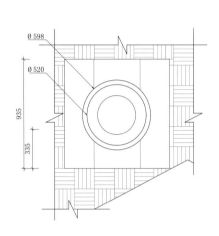

Ø598
Ø520

苏州耦园入口水井平面图
Plan of Well near entrance of Ou Garden, Suzhou

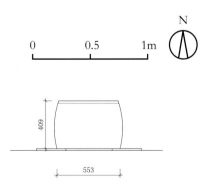

0   0.5   1m   N

苏州耦园入口水井立面图
Elevation of Well near entrance of Ou Garden, Suzhou

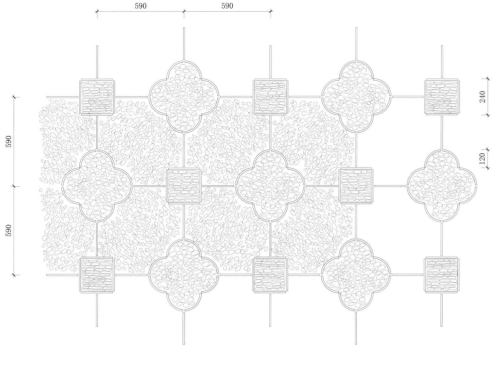

苏州耦园花街铺地（一）大样图
Pavement pattern 1 of Ou Garden, Suzhou

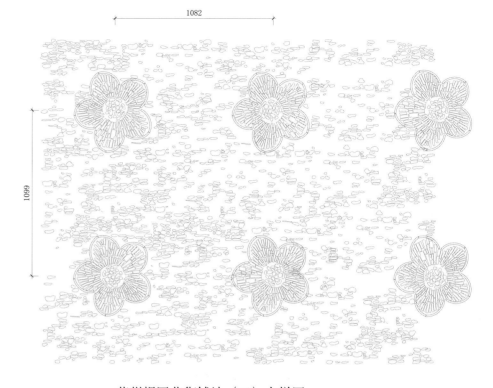

苏州耦园花街铺地（二）大样图
Pavement pattern 2 of Ou Garden, Suzhou

苏州耦园皇道砖铺地大样图
Royal brick pavement of Ou Garden, Suzhou

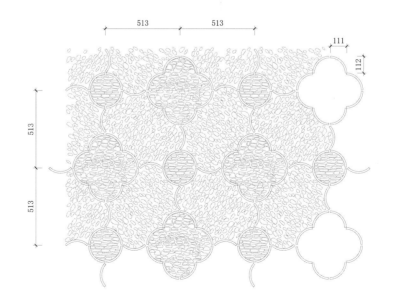

苏州耦园花街铺地（三）大样图
Pavement pattern 3 of Ou Garden, Suzhou

苏州耦园排水口大样图
Drainage of Ou Garden, Suzhou

0   0.5   1m

Liu Garden, Suzhou

Sitting outside Suzhou's western gate, Liu Garden (Lingering Garden) was first built as a garden east of government office from Ming dynasty. It was later rebuilt and named Hanbi (Cold Jade) Residence during the reign of Qing emperor Jiaqing followed by Liu Garden during the reign of emperor Guangxu. A 2-ha expansion to the west, north and east yielded the layout of the garden today. Finally, a comprehensive reconstruction was conducted after the founding of the People's Republic of China.

Liu Garden is comprised of four sections. The midsection, the longest occupied area and location of Hanbi Residence, holds the essence of the garden. This section can be further subdivided into western and eastern halves, the former dominated by hills and water and the latter by buildings and courtyards. In the western half, the hill sits northwest, the pond in the center and the buildings southeast; this arrangement allows sunlight to shine onto the hills and pond—a common practice in Suzhou gardens. The hills are made of earth, with stepped pathways along the pond paved with stacked yellow rockery. Between two northwestern hills flows an inlet, punctuated by a large rock and stone beams. To the pond's east, a series of overlapping buildings sits near Quxi (Meandering Stream) Building while to its south sits Hanbi Cottage, Mingse (Bright Colour) Building and Lvyin Pavilion (Green Shade Pavilion). The arrangement of the buildings amplifies the overlapping heights and offsets. Walking east of Quxi Building, one passes through a number of courtyards, the main one sitting south of Wufengxian (Five-Peak Celestial) Hall. Walking further east, one arrives at Guanyunfeng (Cloud Peak Pinnacle) Courtyard, its layout designed to highlight the rockery pinnacles.

Within a large complex of numerous buildings, Liu Garden boasts a variety of spaces, from small to large, bright to dark, open to closed and high to low; their treatment is precise yet rhythmic. Each courtyard is imbued with its own character and the scene with variety and layers. Most significantly, its halls are amongst the most grand yet elegant ones across Suzhou.

苏州留园

留园在苏州阊门外，创建于明代，原为太仆寺徐泰时的东园。清嘉庆初重修，改称『寒碧庄』（亦称『寒碧山庄』），光绪时改名『留园』，扩建西、北、东三区，成为现有规模，面积约2公顷。中华人民共和国成立后全面修复。

全园大致分四部分，中部寒碧庄经营最久，乃全园精华所在，分东、西两区。西区以山池为主，东区以建筑庭院为主。山池区西北为山，中为池，东南为建筑，使山池主景置于受阳一面，是苏州大型古典园林常例。山为土筑，叠石（黄石为主）为池岸蹬道，西北两山间有水涧，涧口有石矶，架石梁。池东曲溪楼一带重楼杰出，池南有涵碧山房、明瑟楼、绿荫轩等建筑，造型多变，高低错落。曲溪楼东去庭院数区，以主厅五峰仙馆为中心设庭院。再东为冠云峰庭院，布局形式专为突出此峰。

留园规模大，建筑多，空间大小、明暗、开合、高低参差对比，处理精湛，富节奏感。各庭院皆有特色，园景富变化与层次。厅堂在苏州诸园中亦最为宽敞华丽。

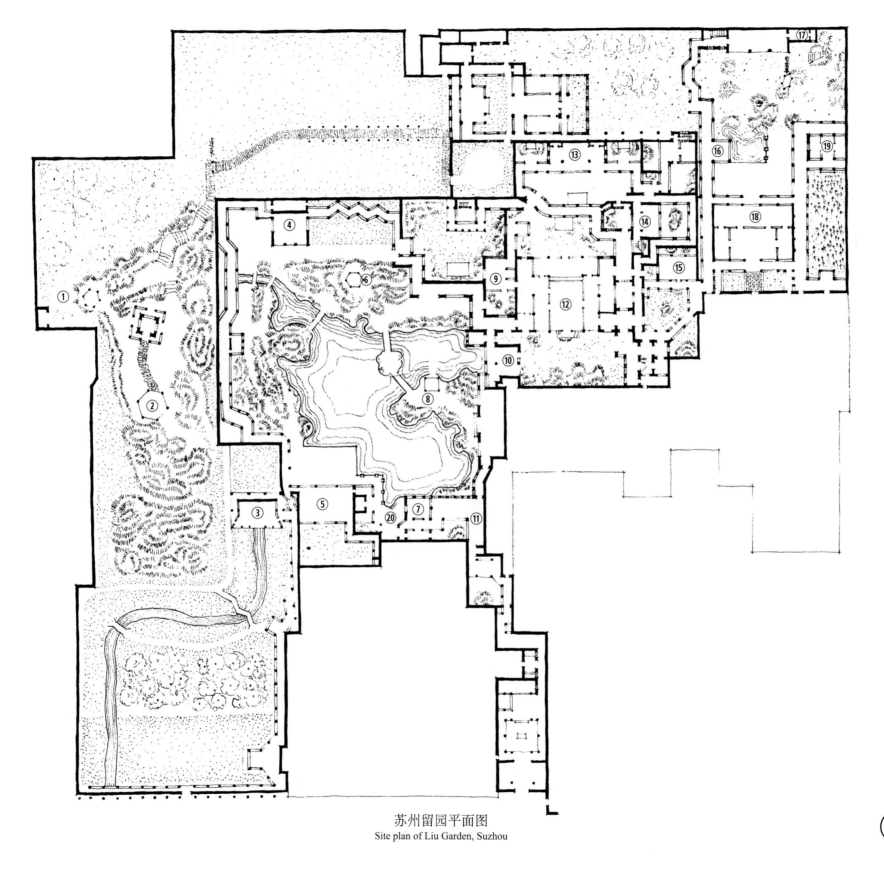

① 舒啸亭 Shuxiao Pavilion
② 至乐亭 Zhile Pavilion
③ 活泼坡地 Huopo podi
④ 半野草堂 Banye Thatched Hall
⑤ 涵碧山房 Hanbi Cottage
⑥ 可亭 Ke Pavilion
⑦ 绿荫轩 Lvyin Pavilion
⑧ 濠濮亭 Haopu Pavilion
⑨ 汲古得绠处 Jigu degengchu
⑩ 清风池馆 Qingfeng chiguan
⑪ 古木交柯 Gumu jiaoke
⑫ 五峰仙馆 Wufengxian Hall
⑬ 花好月圆人寿 Huahao yueyuan renshou
⑭ 还我读书处 Huanwo dushu Study
⑮ 揖峰轩 Yifengxuan
⑯ 冠云台 Guanyun Platform
⑰ 冠云楼 Guanyun Building
⑱ 林泉耆硕之馆 Linquan qishuo Hall
⑲ 待云庵 Daiyun Hut
⑳ 明瑟楼 Mingse Building

苏州留园平面图
Site plan of Liu Garden, Suzhou

N

0   10   20m

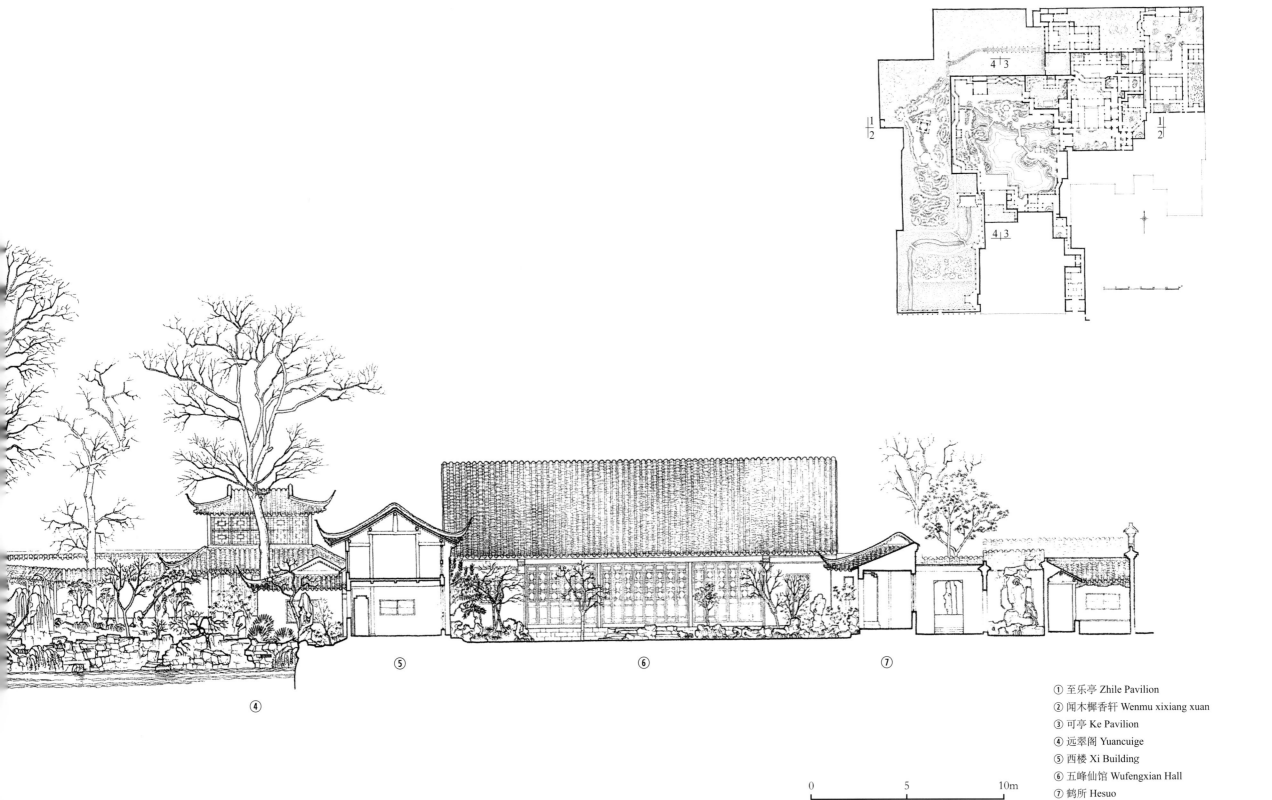

① 至乐亭 Zhile Pavilion
② 闻木樨香轩 Wenmu xixiang xuan
③ 可亭 Ke Pavilion
④ 远翠阁 Yuancuige
⑤ 西楼 Xi Building
⑥ 五峰仙馆 Wufengxian Hall
⑦ 鹤所 Hesuo

0   5   10m

① ② ③

苏州留园 1-1 剖面图
Section 1-1 of Liu Garden, Suzhou

⑤　　　　　　⑥　　　　　　⑦　　　　　　　　　　　　　　　⑧

① 鹤所 Hesuo
② 西楼 Xi Building
③ 濠濮亭 Haopu Pavilion
④ 绿荫 Lvyin Pavilion
⑤ 明翠楼 Mingcuilou
⑥ 涵碧山房 Hanbi shanfang
⑦ 闻木樨香轩 Wenmu xixiang xuan
⑧ 舒啸亭 Shuxiaoting

0　　　　　5　　　　　10m

苏州留园 3-3 剖面图
Section 3-3 of Liu Garden, Suzhou

① ② ③ ④ ⑤ ⑥ ⑦ ⑧

0　　　5　　　10m

① 远翠阁 Yuancuige
② 汲古得绠处 Jigu degeng chu
③ 五峰仙馆 Wufengxian guan
④ 清风池馆 Qingfengchi guan
⑤ 西楼 West Building
⑥ 濠濮亭 Haopu ting
⑦ 曲溪楼 Quxilou
⑧ 涵碧山房 Hanbi shanfang

苏州留园 4-4 剖面图
Section 4-4 of Liu Garden, Suzhou

0   5   10m

① 涵碧山房 Hanbi shanfang
② 闻木樨香轩 Wenmu xixiang xuan
③ 可亭 Keting

Gumu jiaoke (Ancient Tree Foilage) courtyard is located across Lingering Garden's main hill and pond. It serves as a "prologue" to a visitor's journey into the garden.

One reaches Gumu jiaoke courtyard after traversing through a narrow, winding gallery from the entrance. Upon arriving, a row of perforated openings hints at the pond and hills ahead; turning west, one arrives at Lvyin Pavilion– a temporary stop revealing the garden scenery in full openness. This buildup of visual anticipation is a technique called yuyang xianyang (low key before high key).

Huabu xiaozhu (Stride Corridor) is a smaller, adjacent courtyard scene behind Lvyinxuan, a contrast to the wider scenes ahead.

Both courtyards are concisely arranged with flower beds made from bricks or lake rockery planted with bamboos and an ancient tree. Overgrown ivy covers the wall behind—breathing life into the limited space.

的生气。

虎蔓于粉墙，以破墙面单调之感而增加小院

石花台，台上有古树及天竹少许，并以爬墙

两院景致都较简洁，仅设砖花台或湖

正面开朗的景象形成对比。

「华步小筑」是绿荫轩背后的小院，与

手法妙用。

然开朗，山池分外明丽灿烂，是欲扬先抑的

到园中山池景物；向西转至绿荫轩，顿时豁

达「古木交柯」，透过墙上漏窗，可隐约看

由园门而进，经一段曲折晦暗的曲廊而

景区，是入口「序幕」的组成部分。

「古木交柯」庭院位于留园中部山池主

古木交柯与华步小筑
Gumu jiaoke and Huabu xiaozhu

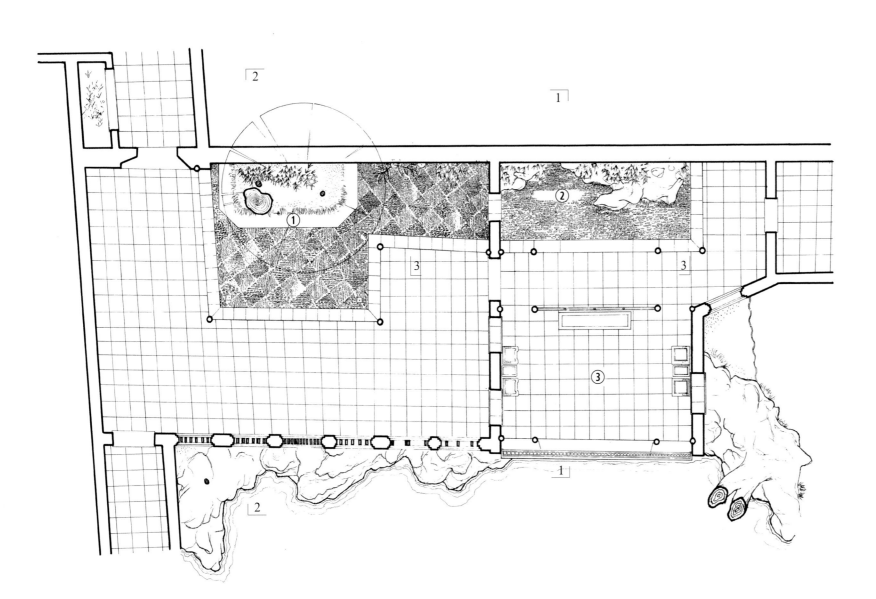

① 古木交柯 Gumu jiaoke
② 华步小筑 Huabu xiaozhu
③ 绿荫轩 Lvyin Pavilion

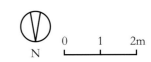

N   0   1   2m

古木交柯与华步小筑平面图
Site plan of Gumu jiaoke and Huabu xiaozhu

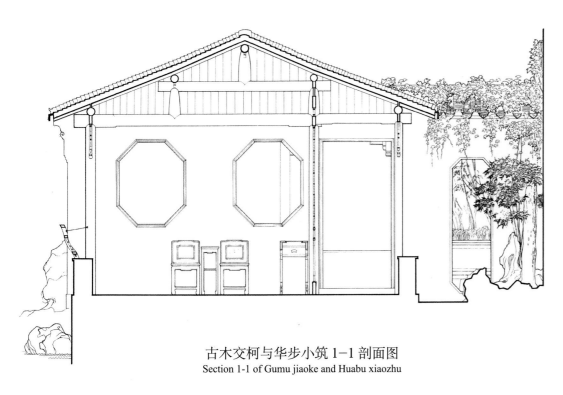

古木交柯与华步小筑 1-1 剖面图
Section 1-1 of Gumu jiaoke and Huabu xiaozhu

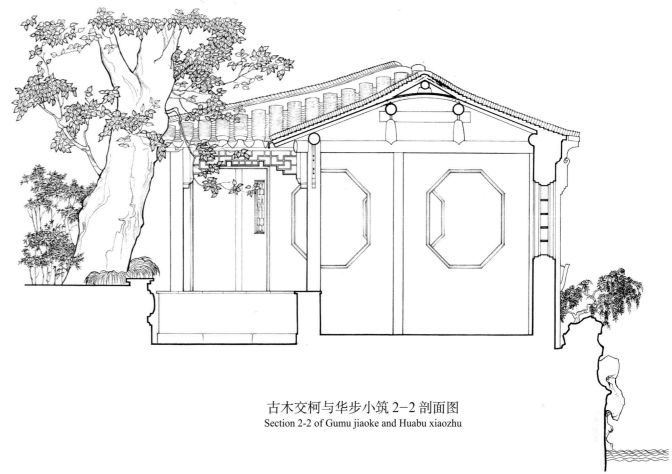

古木交柯与华步小筑 2-2 剖面图
Section 2-2 of Gumu jiaoke and Huabu xiaozhu

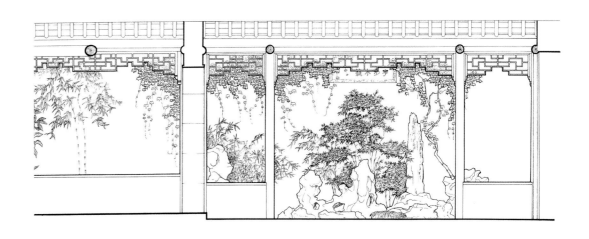

古木交柯与华步小筑 3-3 剖面图
Section 3-3 of Gumu jiaoke and Huabu xiaozhu

0   1   2m

古木交柯与华步小筑剖透视图
Sectional perspective of Gumu jiaoke and Huabu xiaozhu

五峰仙馆前后院
Wufengxian Hall Courtyards

Wufengxian (Five-peaks Celestial) Hall is the main building of Lingering Garden. It was used by its former owners as the garden's main event venue. Consequently, both its front and back courtyards are adorned with impressive sceneries.

The front courtyard's scene is centered on a rockery hill with five peaks; when viewed from the hall, the scene appears much like a scroll painting. To the yard's east stands Hesuo (Crane Pavilion), an auxiliary space with intricate lattice windows. To the west is a two-story building accessible via external steps. Since both the hall and the rockeries are large in scale, the courtyard is rather crammed.

The back courtyard is circled by an undulating gallery from north to east. There is a flower bed in the yard made from lake rockery, on which stands several rock pinnacles, accompanied by maple, pine, bamboo, camellia and other vegetation. These constitute the northern scene of Wufengxian Hall.

五峰仙馆是留园主厅，亦为旧园主的主要活动场所，前后庭院景物丰富。

前院主景为湖石假山，上掇五峰，由厅内观赏，此厅山似一幅长卷画。庭院东面有附属建筑「鹤所」，透空小巧，窗格精致。西面为楼，可由室外楼山拾级而上。由于大厅及院内假山体量较大，所以院内稍有局促之感。

后院东、北二面有曲廊环绕；院中以湖石筑花台，峙石峰数块，配植槭、松、竹、山茶等花木，形成五峰仙馆北面的景观。

① 汲古得绠处 Jigu degeng chu
② 五峰仙馆 Wufengxian guan
③ 鹤所 Hesuo
④ 清风池馆 Qingfengchi guan
⑤ 西楼 West building

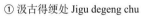

五峰仙馆前后院平面图
Site plan of Wufengxian Hall Courtyards

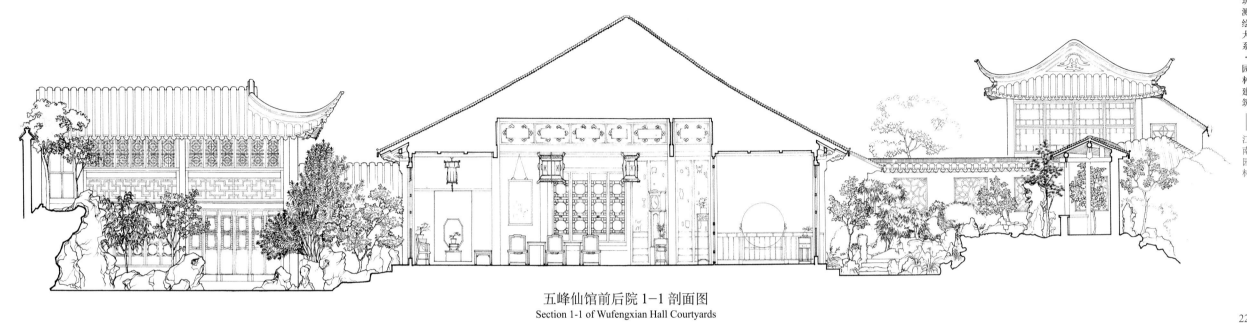

五峰仙馆前后院 1-1 剖面图
Section 1-1 of Wufengxian Hall Courtyards

五峰仙馆前后院 2-2 剖面图
Section 2-2 of Wufengxian Hall Courtyards

0　1　2m

五峰仙馆前后院剖透视图
Sectional perspective of Wufengxian Hall Courtyards

石林小院
Shilin Courtyard

Shilin (Stone Forest) Courtyard is a companion space with several rockery scenes for Wufengxian Hall.

Yifengxuan (Bowing Hill Pavilion) is the main building of the courtyard. It is three-rooms deep, each differing in width, which lends it flexibility in use. Directly opposite sits Shilin Pavilion and the west with a recessed pavilion named Jingzhongguan (Secluded Spot)—both connected by the encircling gallery. In front of Yifengxuan lies the primary scene: a flower bed made from lake rockery and adorned by rockery peaks and mudan (tree peony) shrubs. Yifengxuan's north and west, as well as Shilin Pavilion's three sides, all faces a smaller courtyard—each filled with plants, flowers and rockeries. The use of galleries and pavilions, frequented by punched openings, gates, and long windows, ties the spaces together by blurring the boundary between inside and outside. Although compact, the courtyard is immersed in a multitude of scenes.

The buildings within are small yet refined; the plants are trimmed without appearing overgrown. Most windows and doors were designed towards the rockery pinnacle, accompanied by bamboos, Japanese banana, wisteria, and other plants to put together a pristine, secluded scene.

石林小院是配合大厅五峰仙馆使用的一组庭院，院中造景以石峰为主题。

主建筑揖峰轩三开间，各间面阔不等，布置自由。正对此轩有石林小屋，西侧有半亭「静中观」。曲廊周匝联通。轩前筑湖石花台，上置石峰，植牡丹，成为此轩的主要对景。轩北、西二侧及石林小屋三面又各有小院围绕，院内或植花木，或峙石峰。建筑物多用空廊敞亭及空窗、洞门、长窗，使各院空间贯通，室内外连成一片：院落虽小，层次颇多。

建筑体量与装修小巧玲珑，花木着意修剪，不致蔓衍，与小院空间尺度相协调。门窗框的对景大多以峰石为主，配以竹丛、芭蕉、紫藤等，构成许多清越优美的小景。

228

① 揖峰轩 Yifengxuan
② 鹤所 Hesuo
③ 静中观 Jingzhongguan
④ 石林小屋 Shilin Pavilion

罗汉松
芭蕉
夹竹桃
绣球花
绣球花
紫藤
芭蕉
椿

N   0   1   2m

石林小院平面图
Site plan of Shilin Courtyard

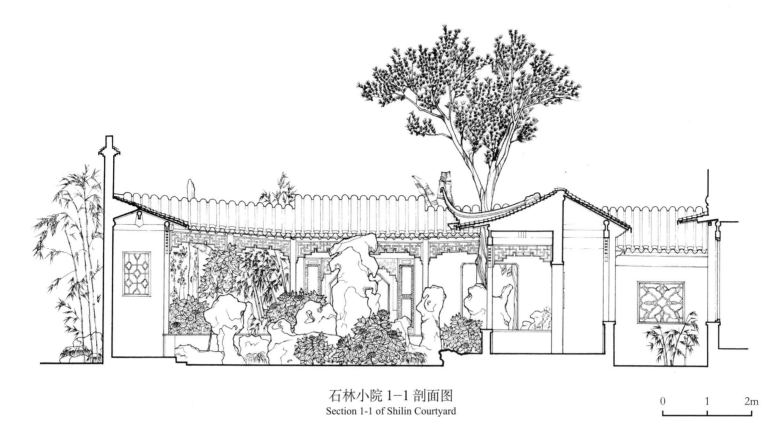

石林小院 1-1 剖面图
Section 1-1 of Shilin Courtyard

0   1   2m

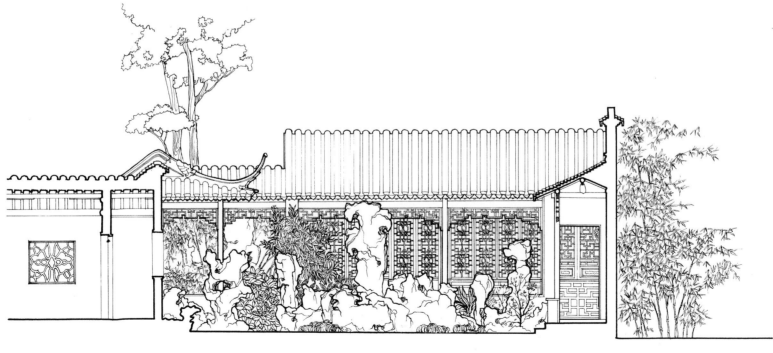

石林小院 2-2 剖面图
Section 2-2 of Shilin Courtyard

0   1   2m

石林小院 3—3 剖面图
Section 3-3 of Shilin Courtyard

0　　1　　2m

石林小院剖透视图
Sectional perspective of Shilin Courtyard

冠云峰庭院
Guanyunfeng Courtyard

Guanyunfeng (Cloud Peak Pinnacle) Courtyard, built during the reign of Qing emperor Guangxu, is a spacious area with tall and open buildings. Rockery pinnacles were erected before the buildings, whereas the buildings were built subsequently around the scene.

The main building, Linquan qishuo (Forest Wisdom) Hall, is a typical yuanyangting (back-to-back hall) directly facing the rockery range to its north. Guanyunfeng, the tallest rockery, is 6.7 m tall and 20 m away from the hall—an viewing distance optimized for appreciating its height. Robust yet elegant, the rockery exposes its natural texture and eroded holes, adorned beneath by a waterlily pond. Directly east of Guanyunfeng sits a hexagon pavilion; it is flanked by Guanyun Pavilion and Zhuyun Hut. Xiuyunfeng, Ruiyunfeng, and other smaller rockeries combine to form a scene of lofty mountain range; other plants and rockeries were kept short or placed aside in keeping with the visual hierarchy.

冠云峰庭院建于清光绪年间，面积较大，建筑物也较高敞。此院先有峰，后建屋，屋为峰设。

主体建筑为『林泉耆硕之馆』，典型的鸳鸯厅，其北面正对冠云峰。冠云峰高6.7米，峰与厅相距约20米，景高与视距的比例较为合适。峰本身挺拔俏丽、纹理自然、孔洞合宜，又前凿小池植睡莲，以水衬石；峰东侧陪衬建有六角亭，峰左、峰右又有冠云台、伫云庵作辅翼，并分置岫云峰与瑞云峰，形成峰峦起伏。庭中花木多低矮，『岫云』『瑞云』二峰也退居两侧，皆为突出主峰。

232

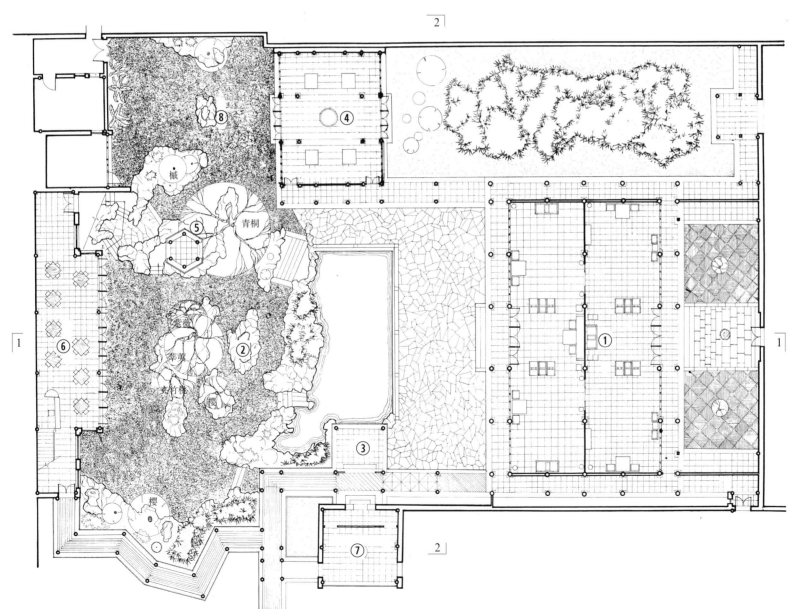

① 林泉耆硕之馆 Linquan qishuo Hall
② 冠云峰 Guanyunfeng
③ 冠云台 Guanyun Platform
④ 待云庵 Daiyun Hut
⑤ 冠云亭 Guanyun Pavilion
⑥ 冠云楼 Guanyun Building
⑦ 佳晴喜雨快雪之亭 Jiaqingxiyukuaixuezhiting
⑧ 瑞云峰 Ruiyunfeng

N 0 1 2 3m

冠云峰庭院平面图
Site plan of Guanyunfeng Courtyard

233

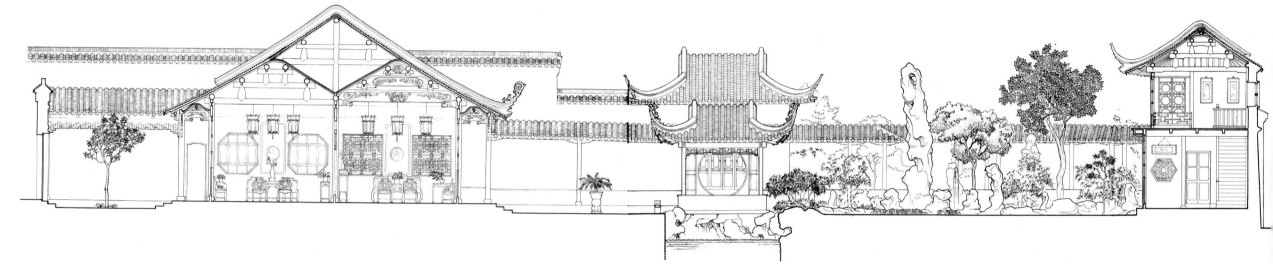

冠云峰庭院 1—1 剖面图

Section 1-1 of Guanyunfeng Courtyard

冠云峰庭院 2—2 剖面图

Section 2-2 of Guanyunfeng Courtyard

0　1　2m

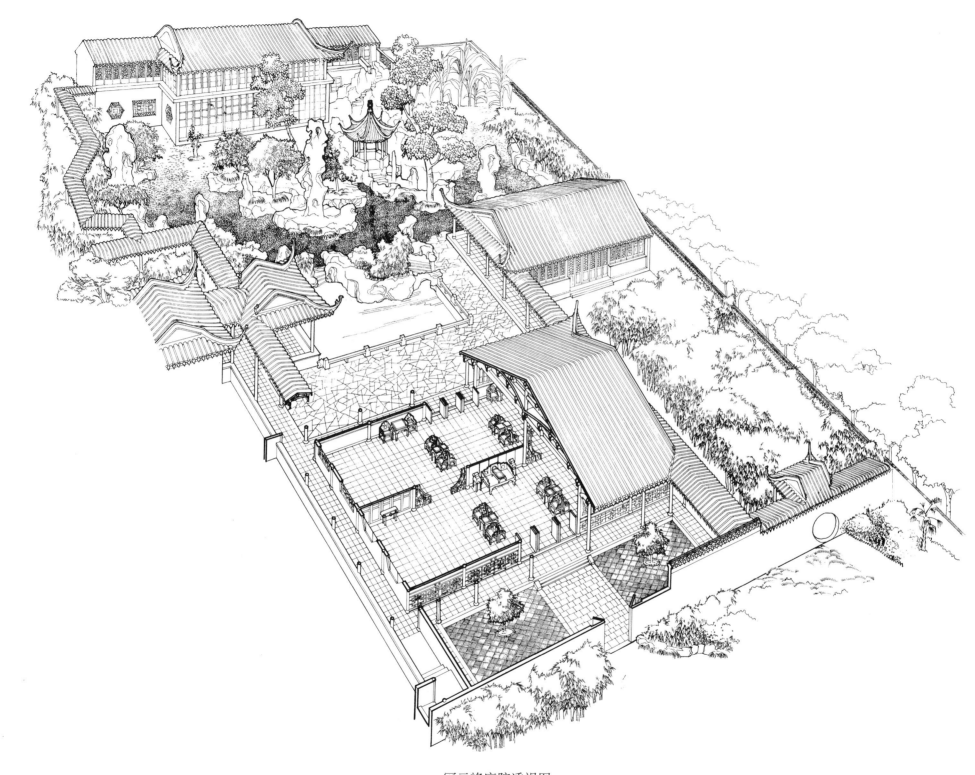

冠云峰庭院透视图
Perspective of Guanyunfeng Courtyard

冠云峰庭院剖透视图
Sectional perspective of Guanyunfeng Courtyard

Jichang Garden, Wuxi

Jichang (Relax Amidst) Garden sits at the eastern foothill of Huishan, the western suburbs of Wuxi city. It was first a monk's residence from Yuan dynasty, then later converted to a garden named Fenggu xingwo, or Nest in Phoenix Valley by a senior military officer in Ming dynasty. It was renamed Jichang Garden during the reign of Ming emperor Wanli, referencing a poem by WANG Xizhi which wrote "to attain wisdom and joy, relax amidst the mountains". In early Qing dynasty, the garden's owner invited two famed gardeners, ZHANG Lian and ZHANG Shi, to renovate the garden. In 1952, the owner's descendants decided to donate the garden to the public, after which the Wuxi municipal government conducted extensive restoration—returning it to its original state.

Occupying a hectare in area, the garden layout follows the terrains from north to south; its main scenes, arranged along the same north-south axis, consist of the pond, earthen hill and buildings. The main building, Jiashu (Splendid Tree) Hall, is positioned at the northmost end of the garden. From here, a southward view captures the garden's most beautiful scene: a surreal, secluded realm of mountains and streams draped in lush greenery. The original entrance was located in the garden's southern tip near Huishan Monastery, allowing longer walks northward to build anticipation. Due to various reasons, the entrance was relocated—now too close to the garden's main scenery.

Most of the garden's buildings were built in late Qing dynasty. Its most significant feat involves borrowing nearby mountains and distant peaks as part of the garden's scenery—incorporating two streams from Huishan as inlets for replenishing the garden's pond. Accompanied by lofty hills and trees, the distant and near are obscured into a continuous backdrop for the garden. While limited in buildings, the lush foliage and misty air lends the garden a naturalistic ambiance.

无锡寄畅园

寄畅园在无锡市西郊东侧的惠山东麓。元代曾是僧舍，明嘉靖初年兵部尚书秦金辟为园，名『凤谷行窝』。后取王羲之『取欢仁智乐，寄畅山水阴』诗中的『寄畅』二字，于万历年间更名『寄畅园』。清初园主曾延请著名造园家张涟（字南垣）及张鉽主持改造。1952 年秦氏后人将园献给国家，无锡市人民政府进行整修保护，逐渐恢复古园风貌。

全园面积约一公顷，布局顺应地形，采取南北走向的分隔，水池、土山及主要观赏建筑都作南北轴向平行布置。在此基础上，主厅『嘉树堂』安排在北面尽头的山水最深处，在此由北南望，得园中最佳景观——山重水复，深林绝涧，如在幽谷，意境极佳。原入口设于南端，靠惠山寺，后因多重原因无奈改筑，现存入口过于逼近主景区，为一憾事。

园中建筑多为晚清后所建。寄畅园充分利用惠山之麓的有利地形，把近山远峰引入园中成为借景，并将惠山二泉之水汇成大池，与土阜乔林构作园中主景，加之建筑稀少，林木葱茏，烟水弥漫，自然风光浓郁。

① 秉礼堂 Bingli Hall
② 含贞斋 Hanzhenzhai
③ 嘉树堂 Jiashu Hall
④ 邻梵阁 Linfange
⑤ 祠堂 Ancestral Hall
⑥ 鹤步滩 Hebutan
⑦ 锦汇漪 Jinhuiyi
⑧ 七星桥 Qixing Bridge
⑨ 涵碧亭 Hanbi Pavilion
⑩ 知鱼槛 Zhiyukan
⑪ 郁盘 yupan
⑫ 碑亭 Stele Pavilion
⑬ 梅亭 Mei Pavilion
⑭ 八音涧 Bayin Stream
⑮ 九狮台 Jiushi Platform

无锡寄畅园平面图
Site plan of Jichang Garden, Wuxi

N   0   10   20m

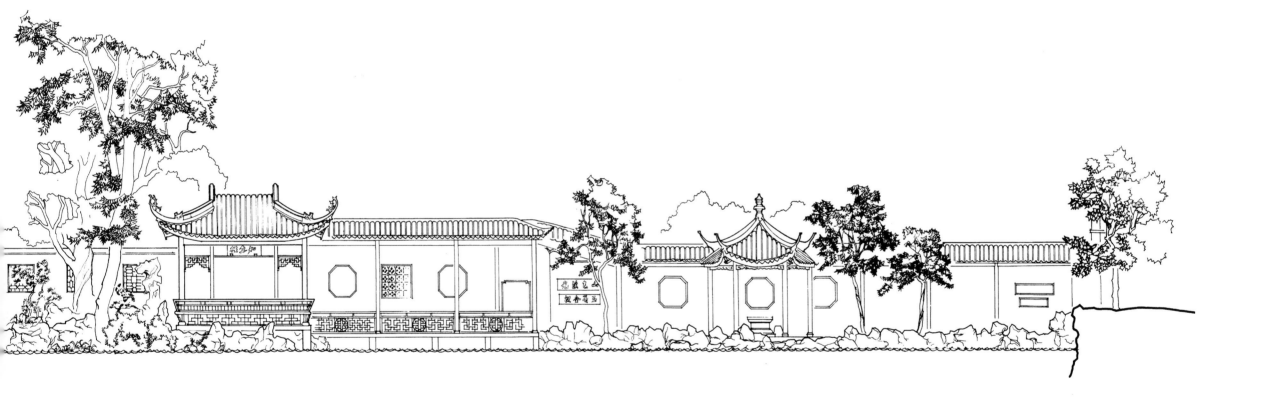

无锡寄畅园 1-1 剖面图
Section 1-1 of Jichang Garden, Wuxi

无锡寄畅园 2-2 剖面图
Section 2-2 of Jichang Garden, Wuxi

0 10 20m

秉礼堂庭院
Bingli Hall Courtyard

Located in the southwestern corner of Jichang Garden, Bingli (Upheld Respect) Hall is the flower hall of Zhenjieci's western courtyard. In terms of craftmanship, it is the most refined building in the garden and a remarkable Qing dynasty relic with a style prevalent of its mid-period. It is three-rooms wide with *guanyindou* (gable walls of curved profile) on either end, enclosing a well-crafted heavy timber structure and fine tileworks within. The rockeries and pond were added later in the 20th century.

Bingli Hall's rectilinear courtyard is centered on a pond encircled by galleries to its east, west, and north. The west gallery terminates on either end in small courtyards—a mirrored scene when viewed from the middle. Each courtyard is dotted by lake rockeries, flowers and trees. These simple techniques, austere layout, and stratified scenes lends the courtyard a secluded atmosphere.

秉礼堂在寄畅园西南隅，是贞节祠的西院花厅，在园中诸建筑中制作最精，有清代盛期风格，为清中叶遗物。建筑三开间硬山，观音兜山墙，木构架用料工整，室内轩及砖细加工精致。假山、池沼为20世纪新建。

秉礼堂庭院基地方整，以水池为中心，东、西、北三面绕以回廊，西廊两端形成小院，点湖石花木，组成回廊的对景。疏院景繁简得体，萧疏简朴，层次较多，自成一区幽僻庭院。

桑树

大叶黄杨

栀子

匍地柏

石榴

罗汉松

紫荆

栀子

火叶榆

瓜子黄杨

桂花

银边黄杨

紫藤

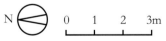

N 0 1 2 3m

秉礼堂庭院平面图
Site plan of Bingli Hall Courtyard

秉礼堂庭院 1-1 剖面图
Section 1-1 of Bingli Hall Courtyard

0 1 2m

秉礼堂庭院 2-2 剖面图
Section 2-2 of Bingli Hall Courtyard

0　1　2m

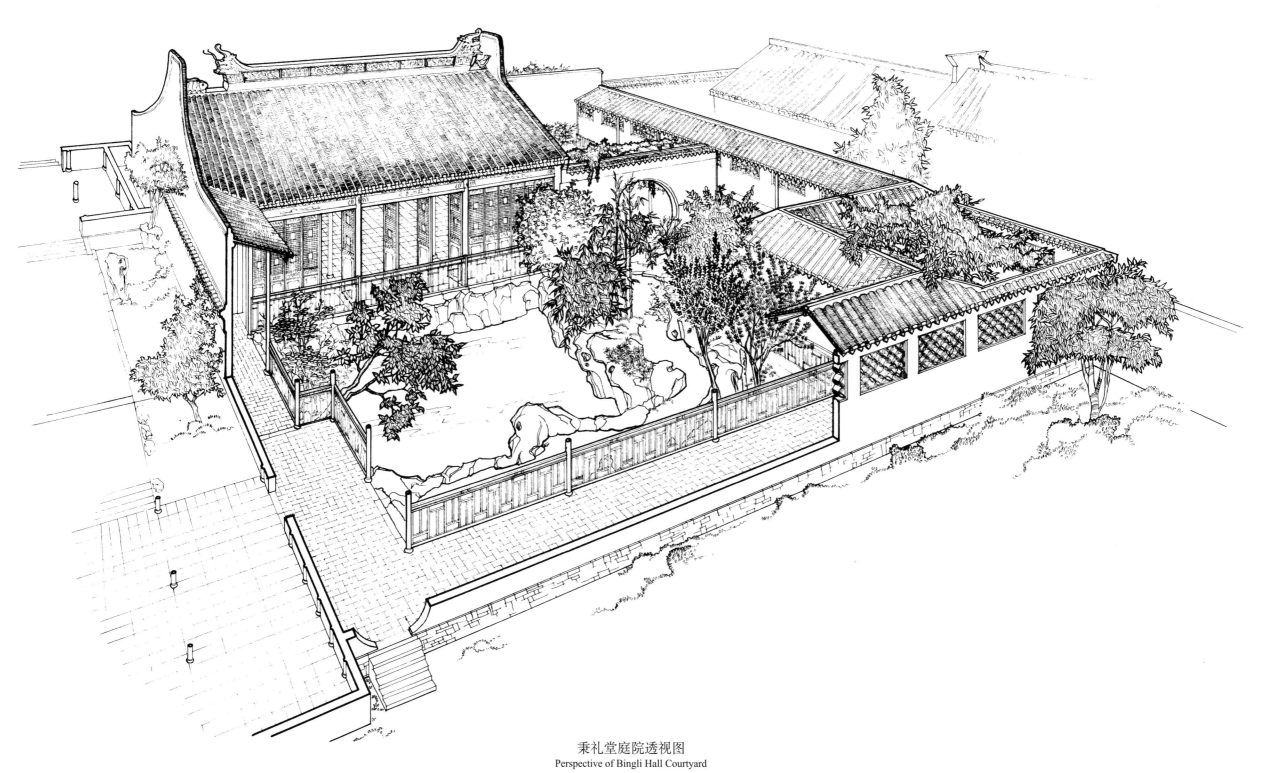

秉礼堂庭院透视图
Perspective of Bingli Hall Courtyard

Tuisi Garden, Wujiang

Tuisi (Retreat and Reflection) Garden is located in Tongli, Wujiang district of Jiangsu province, first built in 1885. It was originally the residential garden of REN Lansheng, a retired Qing government official. The garden sits to the residence's east, covering an area of 2,500 m sq; it is divided into east and west sections.

The eastern section is the garden's primary space, centered on a pond surrounded by rockery hills, pavilions, flowers and trees. Tuisi caotang (Reflection Thatched Hall) is the main building of the section, sitting on the northern banks of the pond. It is a four-sided structure with a platform in front, offering views of surrounding sceneries and the red carps nearby. Along the pond's western bank, a gallery traces along the water, leading to a boat-shaped pavilion midway. Through the gallery's perforated openings, one can also catch glimpses of courtyards in the west. The pond's east is marked by a rockery hill; to its south, a small building and a waterside pavilion named Guyu shengliang (Cooling Rain from Shallow Waters); this elevated pavilion is ascended via stone steps, upon which holds a panoramic view of the garden. On the other side, the garden's western section is comprised of courtyards, the owner's studies and venue for greeting guest. Zuochun wangyue (Gaze upon Spring Moon) Building, at six-bays in width, is the main architecture of the western section. In an courtyard along the western wall sits a small pavilion of three bays, resembling a three-cabin boat; it opens eastward facing the distant mountains, flanked by lattice windows on either sides. To the east stands a flower bed of lake rockery serving as a small backdrop. To the south, Yingbinshi (Greeting Chamber) and Suihanju (Winter Cottage) face Zuochun wangyue Building.

吴江退思园

退思园在江苏吴江市同里镇，始建于清光绪十一年（1885年），是清代凤、颖、六、泗兵备道任兰生的宅园。园在住宅东侧，占地约 2500 平方米，分为东、西两部分。

东部是退思园主体，以水池为中心，环池布列假山、亭阁、花木。『退思草堂』为其主体建筑。堂在池北岸，作四面厅形式，前有平台临水，可周览环池景色，或俯察水中碧藻红鱼。池西有一带曲廊贴水而前与旱船相连，透过廊内漏窗，可隐约窥见西部庭院景物。池东有假山，池南有小楼及临水小轩『菰雨生凉』，由室外石级登小楼，可俯瞰全园。西部是主人读书待客之所，以建筑庭院为主。主体建筑为书楼『坐春望月楼』，六开间。楼前院内依西墙建三间小斋，仿画舫前、中、后三舱之意，于山面开门，两侧仅开和合窗。院东侧近墙处设湖石花台，疏植花卉树木，构成小景。院南侧面对书楼则有『迎宾室』『岁寒居』等斋馆。

吴江退思园屋顶平面图
Roof plan of Tuisi Garden, Wujiang

N

0   5   10m

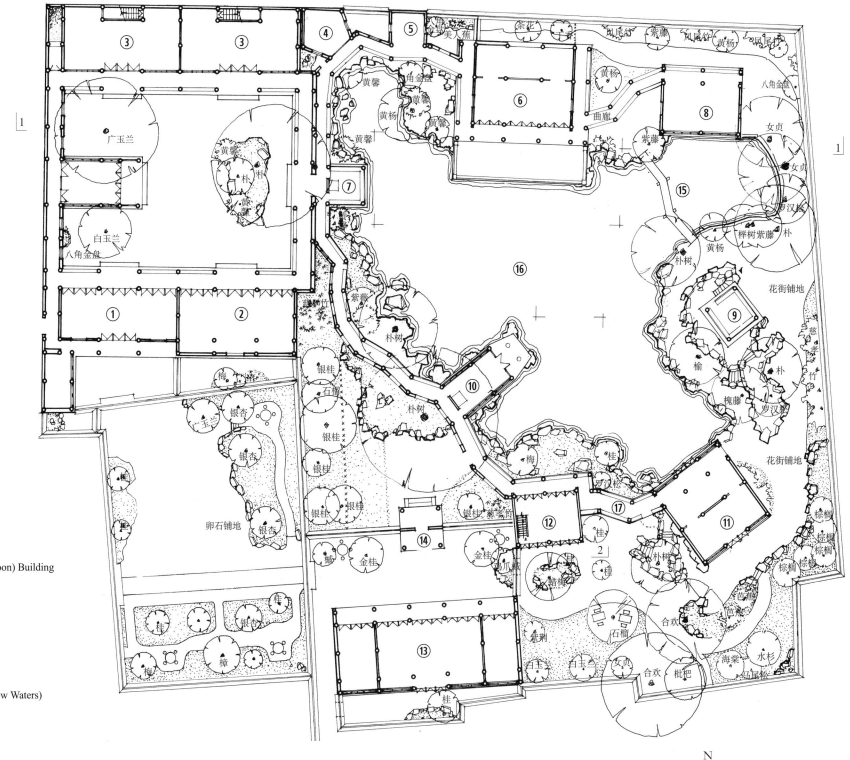

① 岁寒居 Suihanju (Winter Cottage)
② 迎宾室 Yingbinshi (Greeting Chamber)
③ 坐春望月楼 Zuochun wangyue (Gaze upon Spring Moon) Building
④ 揽胜阁 Lanshengge
⑤ 小轩 Small Pavilion
⑥ 退思草堂 Tuisi caotang (Reflection Thatched Hall)
⑦ 水香榭 Shuixiangxie
⑧ 琴房 Music Room.
⑨ 眠云亭 Mianyun Pavilion
⑩ 闹红一舸 Naohongyige
⑪ 菰雨生凉 Guyu shengliang (Cooling Rain from Shallow Waters)
⑫ 辛台 Xin Platform
⑬ 桂花厅 Guihua Hall
⑭ 亭 Pavilion
⑮ 三幽桥 Sanyou Bridge
⑯ 荷花池 Lotus Pond
⑰ 天桥 Foot Bridge

吴江退思园一层平面图
Plan of first floor of Tuisi Garden, Wujiang

N

0   5   10m

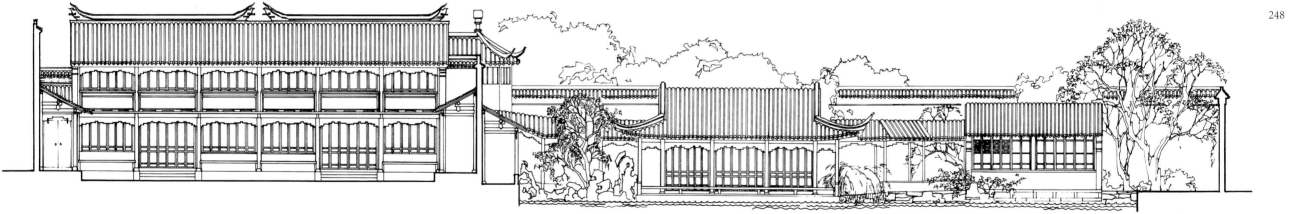

吴江退思园 1-1 剖面图
Section 1-1 of Tuisi Garden, Wujiang

0 1 2 3m

吴江退思园 2-2 剖面图
Section 2-2 of Tuisi Garden, Wujiang

0　1　2　3m

Zhan Garden, Nanjing

Located on Zhanyuan Road in Nanjing, Zhan (Look Upward) Garden was built by XU Pengju, the seventh descendent of XU Da during the reign of Ming emperor Jiajing. The rockery hill sitting in the garden's north is the only original left from Ming dynasty; the rest was done in a restoration effort led by professor LIU Dunzhen in 1960, followed by an expansion in 1988.

The garden is centered on Jingmiao (Quiet contemplation) Hall with hills and waters to its north and south. Stepping in the garden entrance on Zhanyuan Road, one reaches Jingmiao Hall through a long gallery, passing small courtyards and pavilions along the way. As one moves through the various spaces, the scenery unfolds from dark to bright—opening up unto vast landscapes northward and southward. Through perforated openings on either side of the gallery, glimpses of the courtyards and garden leaks into view. To the north of the hall lies a lawn, beyond of which sits the rockery hills and pond from Ming dynasty. Its immediate west is a gently sloped earthen hill accessed by footbridges; stretching from north to south, it sits amidst lush foliage, blocking city buildings from view whilst permeating the air of a wild forest. The northmost rockery range, imposing in scale, is hollowed with caves and looped with pathways. In front, a waterside path leads to a stone footbridge on either end; flanked by rockery on one side, the narrow path meanders between the pond and boulders. These garden techniques were extensively used in Ming dynasty, evoking scenes from a stroll along a narrow path between valleys and streams. To the south of Jingmiao Hall, another rockery hill sits across the pond. This was designed by professor LIU Dunzhen, who meticulously conceived its undulating contours from hills to ravines.

南京瞻园

瞻园在南京瞻园路，建于明嘉靖年间，为徐达七世孙徐鹏举所创。今园中仅北部假山为明代遗物。1960 年由刘敦桢先生主持作整治修复工作，1988 年又进行扩建。

瞻园布局以静妙堂为中心。从瞻园路入园门，由长廊引导，经小庭院、小轩、亭榭而至静妙堂，景面逐步展开，空间由小而大，由暗而明，南北山池都豁然开朗。长廊两侧还可通过漏窗隐约见到小庭院及园中景色。厅堂之北，留有一片草坪，草坪以北即是明代所遗假山、水池。其西有土山一垄，陵阜陂陀，平岗小坂，由南向北延伸，止于池西，山上林木茂密，具有天然山林野趣，同时又将西墙外的建筑物挡于视线之外。池北假山体量较大，中构山洞，山上有磴道盘回。山前临水有小径，由东西二石梁联络两岸。小径一侧依石壁，一面面水池，并伸出石矶漫没水中，略具江河峡间纤路、栈道之意，是明代假山常用手法。静庙堂之南亦有山与之隔池相望，此假山为刘敦桢先生主持设计，对轮廓起伏、丘壑虚实都颇有考虑。

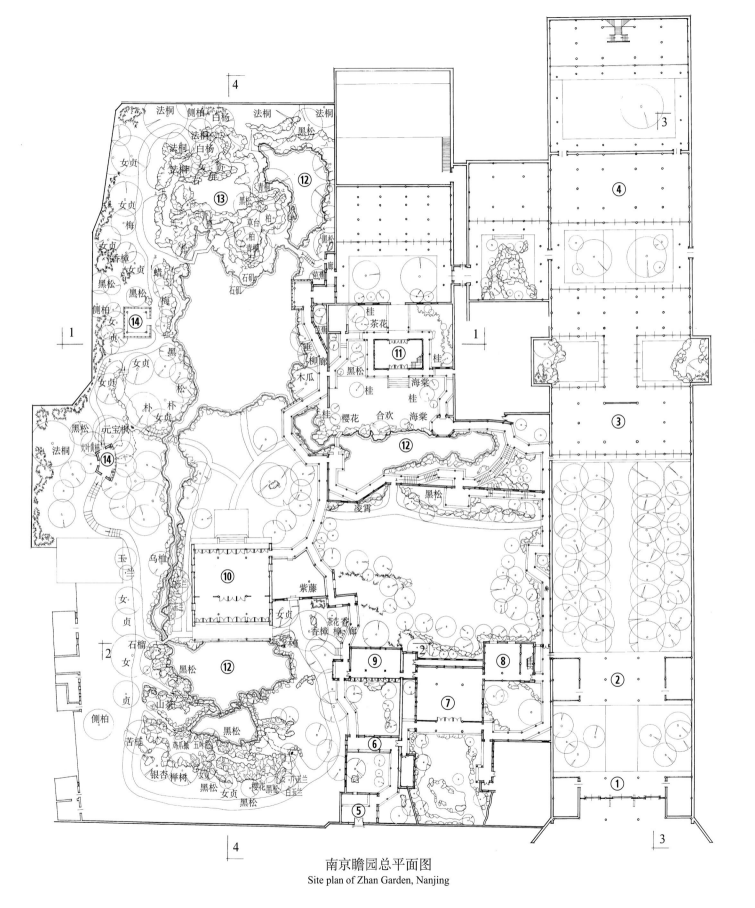

法桐　侧柏　白杨　法桐　法桐
女贞
法桐　白杨　女贞　黑松
栾树　女贞
女贞　⑬
梅
女贞　草台　柏　青枫
山　香樟　石矶
女贞　石矶　黑松
黑松　梅
侧柏　黑松　⑭
女贞
黑　垂柳廊
女贞　松　木瓜
朴　朴　女贞　⑫　樱花　合欢　海棠
黑松　元宝枫　⑭　黑松
法桐　⑫　凌霄
乌桕
玉兰　⑩　紫藤
女贞　女贞
石榴　茶花香　樟廊
女贞　香樟
侧柏　⑫　⑨　②　⑧
黑松　⑦
山楂　黑松　⑥
苦槠
银杏榉树　黑松　⑤
黑松　女贞　樱花黑松

桂　茶花
⑪　桂
黑松　桂　海棠
桂　桂

④　③

⑬　草台 Grass Platform
⑫　水池 Pond

② Instrument Hall
① The Hall

① 大厅 The Hall
② 仪厅 Instrument Hall
③ 大堂 Lobby
④ 玉兰堂 Yulantang
⑤ 门廊 Porch
⑥ 致爽轩 Zhishuangxuan
⑦ 接待室 Reception Room
⑧ 办公 Office
⑨ 花篮厅 Flower Hall
⑩ 静妙堂 Jingmiao Hall
⑪ 一览阁 Yilange
⑫ 水池 Pond
⑬ 草台 Grass Platform
⑭ 亭 Pavilion

南京瞻园总平面图
Site plan of Zhan Garden, Nanjing

N

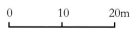

0　10　20m

南京瞻园 1—1 剖面图
Section 1-1 of Zhan Garden, Nanjing

0　　　　5　　　　10m

南京瞻园 2-2 剖面图
Section 2-2 of Zhan Garden, Nanjing

0　　　　　5　　　　　10m

0   5   10m

南京瞻园 3-3 剖面图
Section 3-3 of Zhan Garden, Nanjing

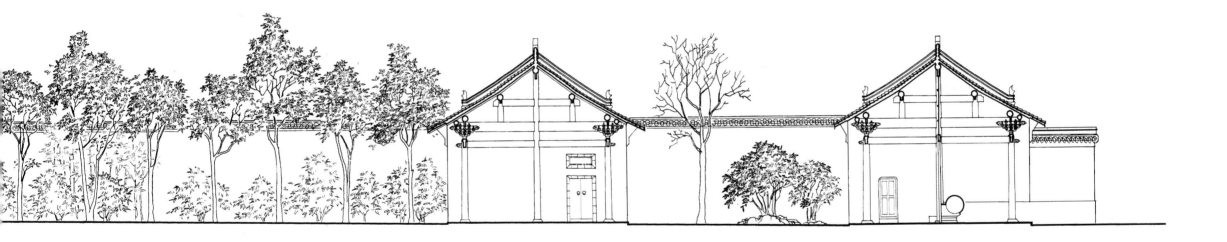

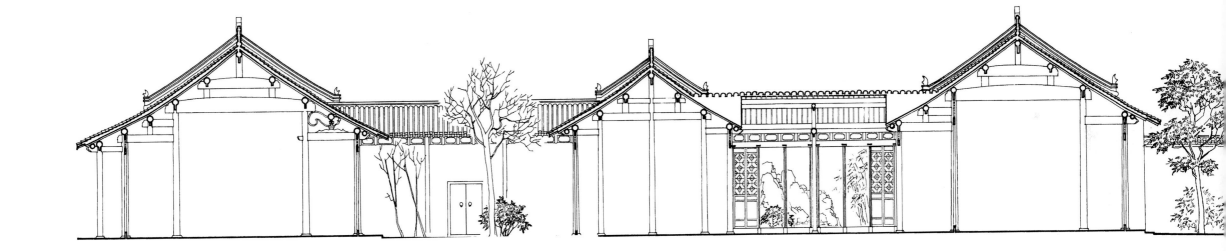

南京瞻园 4-4 剖面图
Section 4-4 of Zhan Garden, Nanjing

Xu Garden, Nanjing

Located on Changjiang Road in Nanjing, Xu (Warmth) Garden was originally a leisure garden attached to the west of governor's office in Qing dynasty. During the Taiping Rebellion, it became the garden attached to HONG Xiuquan's residence and was renamed West Garden. After the rebellion fell, the garden was destroyed. The buildings, halls, pavilions, and rockery hills there today were built later by ZENG Guofan during the reign of emperor Tongzhi.

The garden is laid out along a pond surrounded by pavilions, trees, bamboos and rockeries. Long from north to south, narrow from east to west, the pond is shape much like a bottle. In the middle of the pond stands Yilange (Ripple Pavilion), with a platform out front and a view southward onto the boat-shaped structure Buxizhou. To the pond's east sits Wangfeige (Forgotten Flight Pavilion); to its west, Xijia (Sunset) Building. The four structures constitute the main waterside scene, each serving as a focus from the other's view. Further east of the pond sits Tongyin (Tung Tree Sound) Hall, its front and back each facing a rockery hill. Between the hall and the pond, a *yuanyangting* (hybrid pavilion) sits on granite pavement; its composite pyramidal roof is especially well-crafted. A hexagonal pavilion is also located on a nearby rockery hill.

南京煦园

煦园在南京长江路。清代属两江总督府西花园，太平天国时洪秀全以督署建为天王府，此园为天王西花园。当时都城天京（现南京）失陷，园毁坏殆尽，现存楼、馆、亭、阁及假山多系同治年间曾国藩重建总督衙门时所建。

此园布局以水池为中心，环池布置亭阁、竹树、假山。水池南北长，东西狭，略呈瓶形。池中近北处建一水阁，为『漪澜阁』，阁前设平台，可眺池南水中石舫『不系舟』、两岸楼阁。池东西分别有『忘飞阁』与『夕佳楼』。这四座建筑依托池面，互为对景。水池东侧一区以桐音馆为主要建筑，前后各构湖石假山一座作为此馆对景。馆侧和池水间的陆地有一花岗地，上置鸳鸯亭——套方攒尖双亭，十分别致。另有六角亭一座，位于假山上。

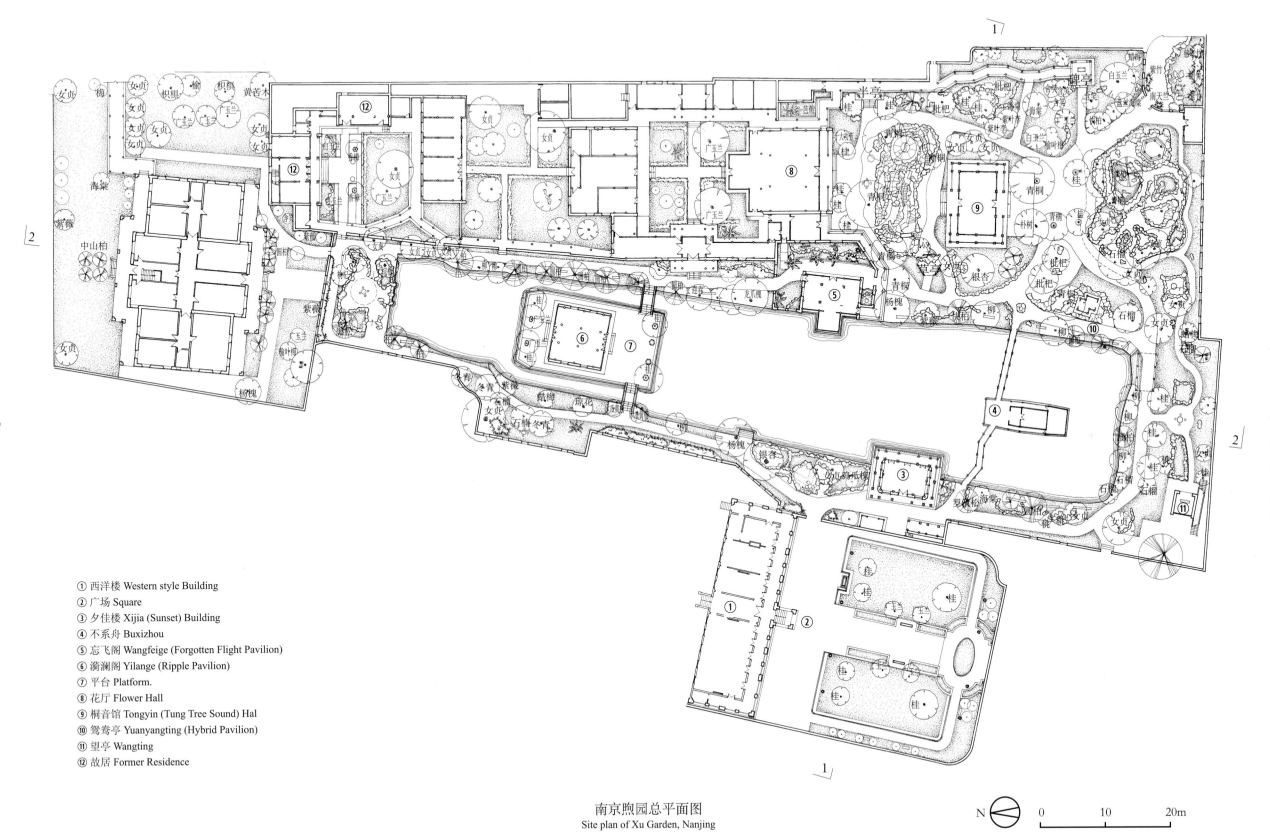

① 西洋楼 Western style Building
② 广场 Square
③ 夕佳楼 Xijia (Sunset) Building
④ 不系舟 Buxizhou
⑤ 忘飞阁 Wangfeige (Forgotten Flight Pavilion)
⑥ 漪澜阁 Yilange (Ripple Pavilion)
⑦ 平台 Platform.
⑧ 花厅 Flower Hall
⑨ 桐音馆 Tongyin (Tung Tree Sound) Hal
⑩ 鸳鸯亭 Yuanyangting (Hybrid Pavilion)
⑪ 望亭 Wangting
⑫ 故居 Former Residence

南京煦园总平面图
Site plan of Xu Garden, Nanjing

N   0   10   20m

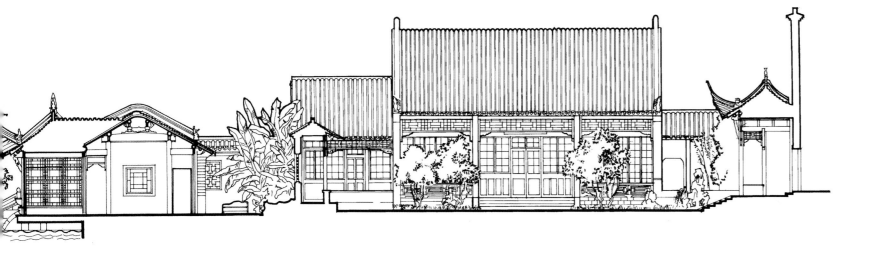

0   5   10m

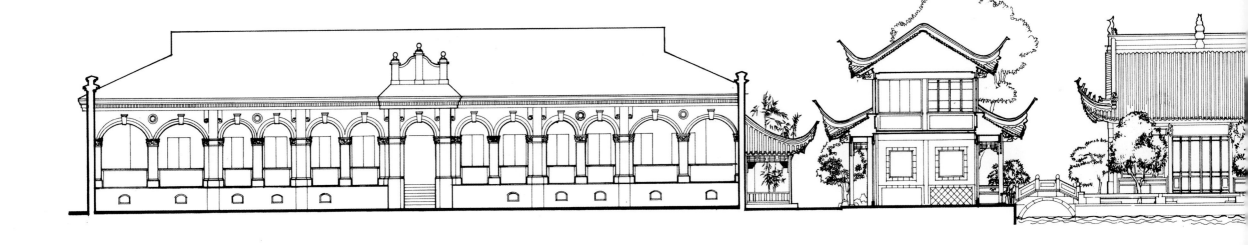

南京煦园 1-1 剖面图
Section 1-1 of Xu Garden, Nanjing

0   5   10m

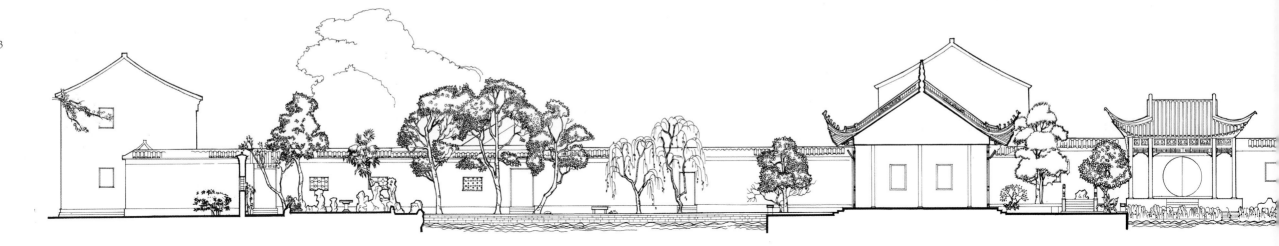

南京煦园 2-2 剖面图
Section 2-2 of Xu Garden, Nanjing

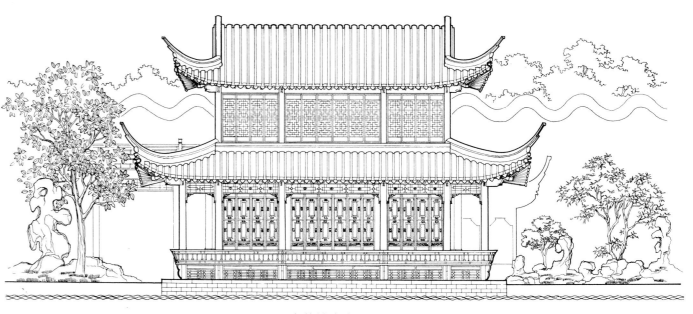

夕佳楼东立面图
East elevation of Xijia Building

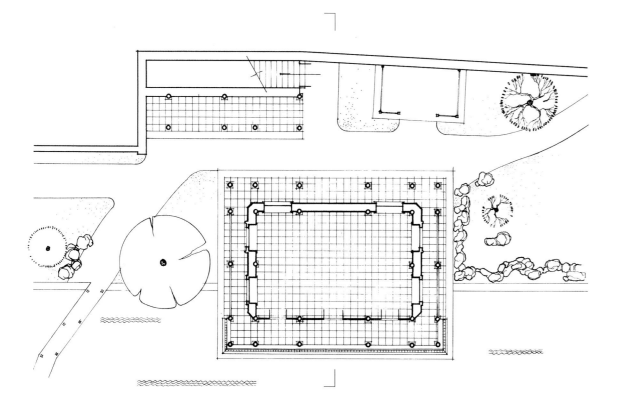

夕佳楼一层平面图
Plan of first floor of Xijia Building

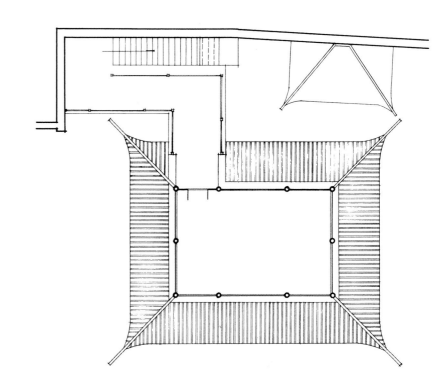

夕佳楼屋顶平面图
Roof plan of Xijia Building

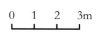

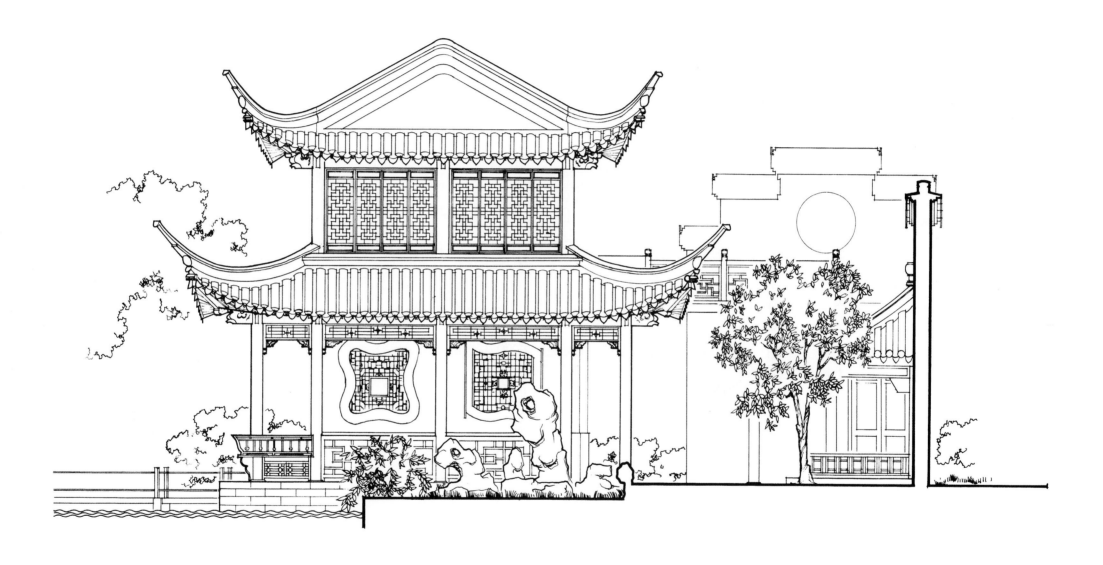

夕佳楼北立面图
North elevation of Xijia Building

0 1 2m

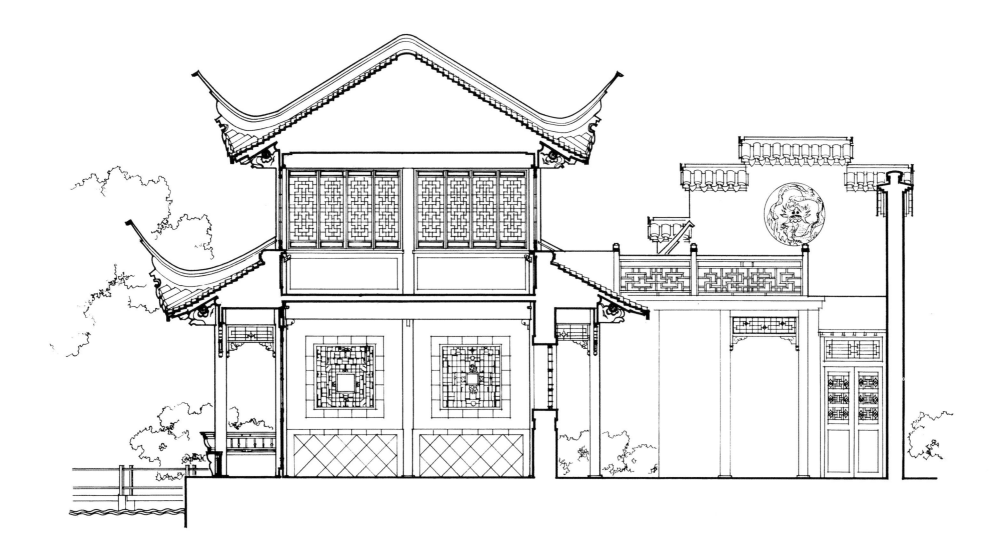

夕佳楼剖面图
Section of Xijia Building

0　　1　　2m

不系舟
Buxizhou

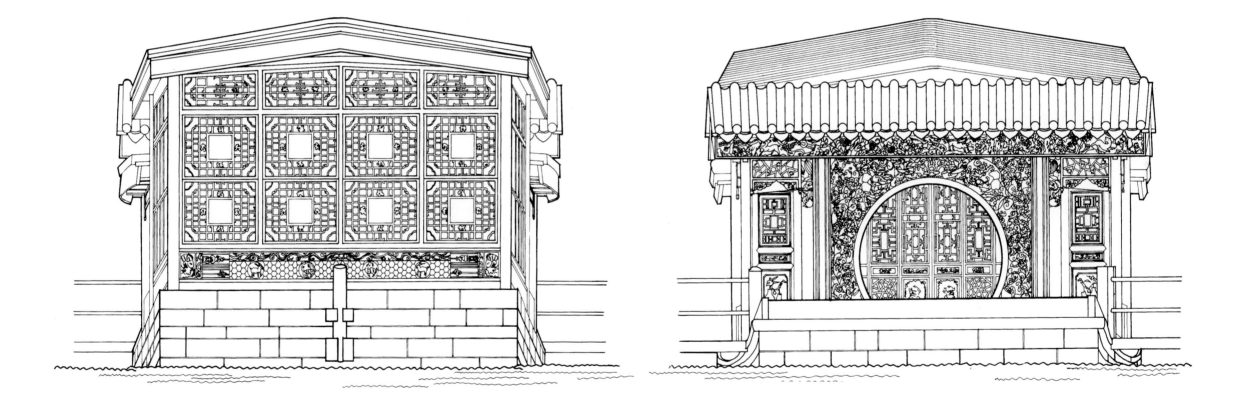

不系舟北立面图
North elevation of Buxizhou

不系舟南立面图
South elevation of Buxizhou

0   1   2m

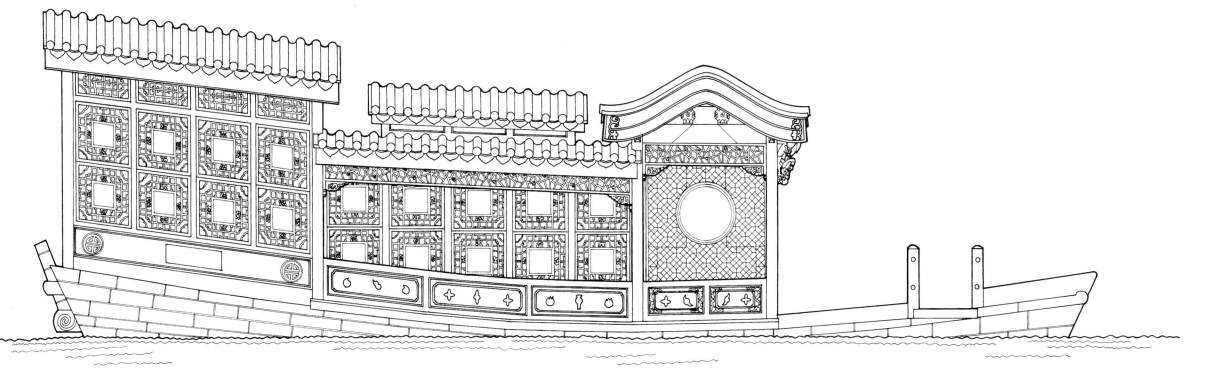

不系舟东立面图
East elevation of Buxizhou

0　　　　1　　　　2m

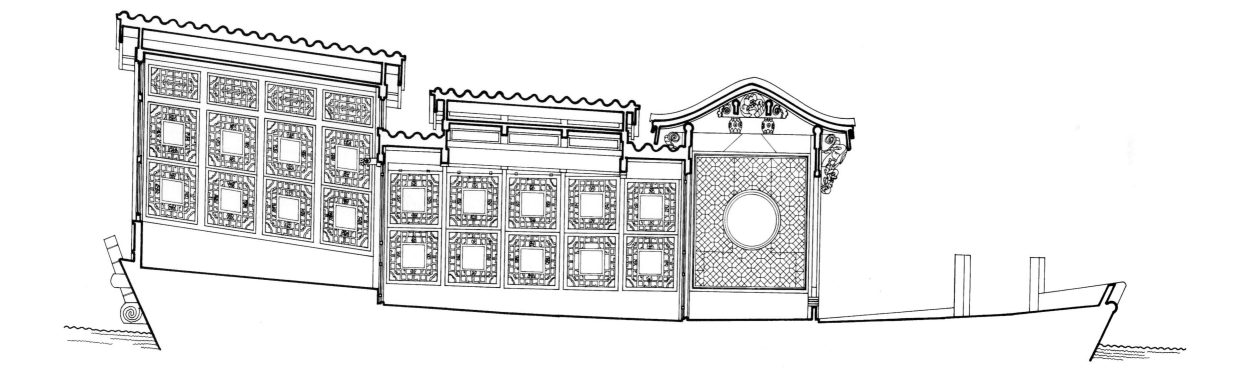

不系舟剖面图
Section of Buxizhou

0   1   2m

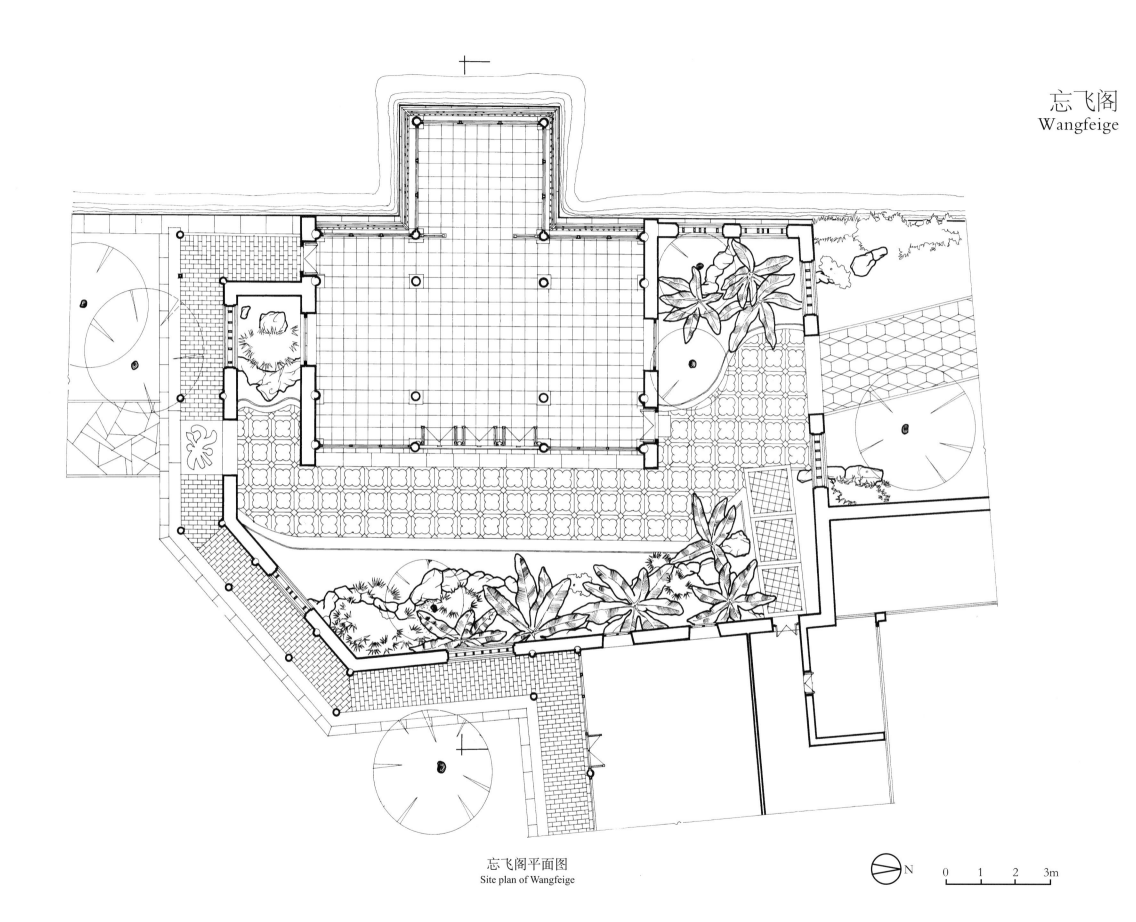

忘飞阁平面图
Site plan of Wangfeige

N   0   1   2   3m

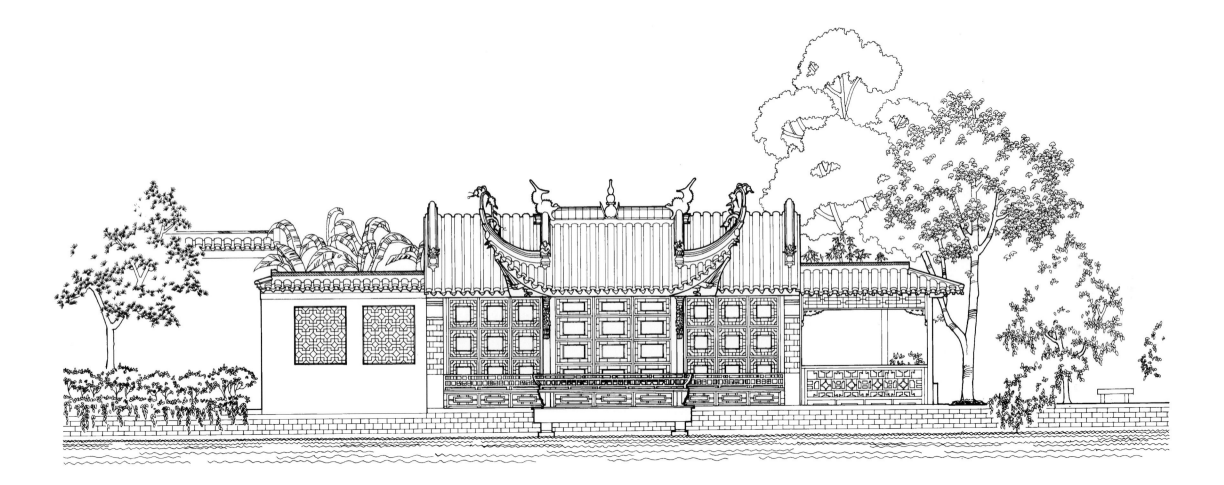

忘飞阁西立面图
West elevation of Wangfeige

0 1 2 3m

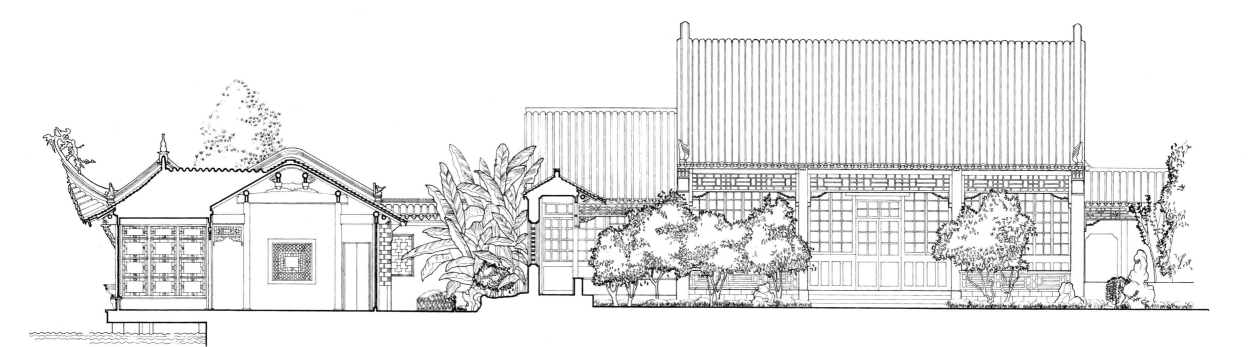

忘飞阁剖面图
Section of Wangfeige

0 1 2 3m

He Garden, Yangzhou

He Garden, first built in 1862, is located on Xuningmen huayuan Lane of Yangzhou. In 1883, the rockery garden south of it named Pianshi shanfang (Stone Sheet Residence) was added to the premise. Overtime, frequent changes of owners led to its eventual deterioration. In 1979, it was finally handed to landscape authorities for restoration. Subsequently, the rockery hills from Pianshi shanfang and its Nanmu Hall from Ming dynasty were also restored as part of Heyuan. It is a major cultural site protected at the national level.

Positioned behind the residence, He Garden covers an area of 0.4 ha and is split into eastern and western sections. The eastern section is centered on a flower hall surrounded by rockeries, pond, and pavilions. Conversely, the western section is centered on a pond and rockeries, circled by buildings and two-storied galleries. The former is designed mainly for indoor activities and the latter for outdoors.

扬州何园

何园在扬州徐凝门花园巷，建于清同治元年（1862 年），光绪九年（1883 年）将吴氏片石山房并入园内。后经历住宅易主、园林残败，1979 年由园林部门收回、整修开放，随后将『片石山房』所遗假山及明构楠木厅整修归于园内。现为全国重点文物保护单位。

园位于住宅后部，面积约 0.4 公顷，分东西两区。东区以花厅为中心，四周配山池、亭廊；西区以山水为主题，四面环房屋。前者以室内活动为主要内容，后者以室外游观为构思依据。

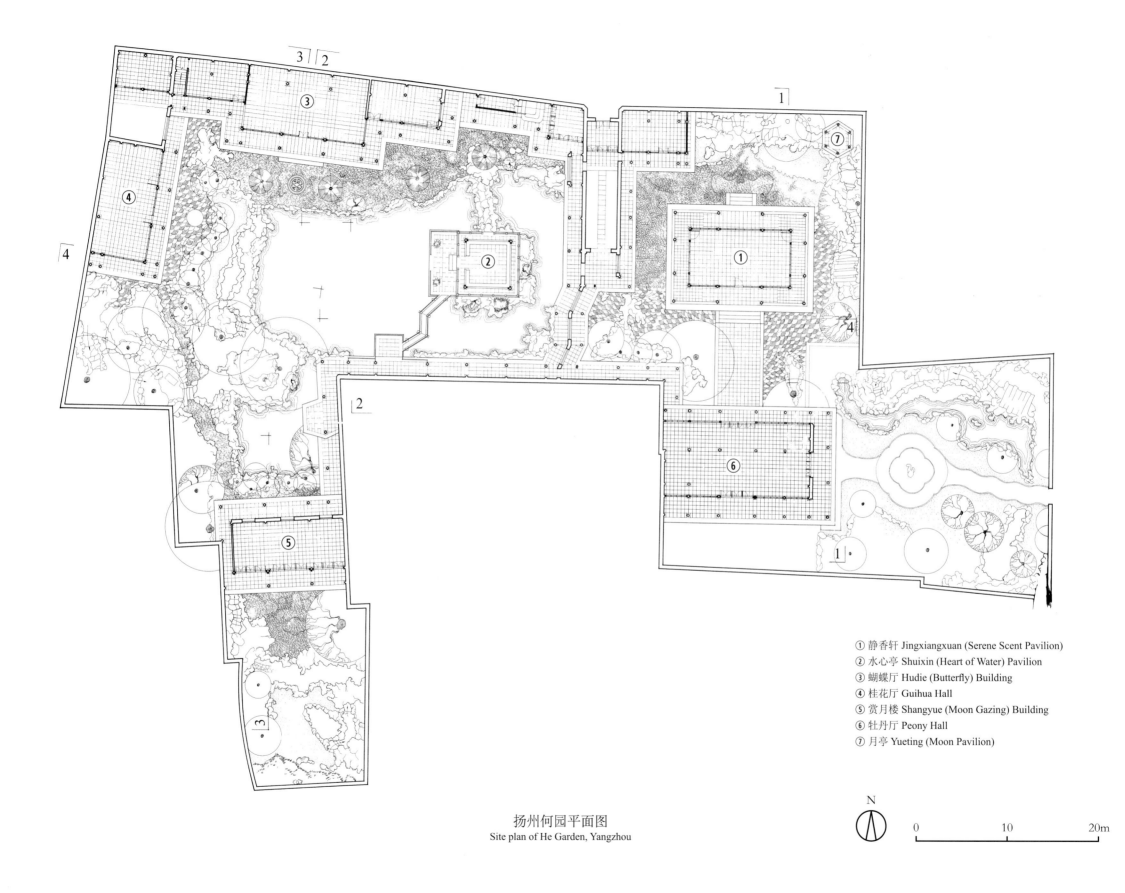

① 静香轩 Jingxiangxuan (Serene Scent Pavilion)
② 水心亭 Shuixin (Heart of Water) Pavilion
③ 蝴蝶厅 Hudie (Butterfly) Building
④ 桂花厅 Guihua Hall
⑤ 赏月楼 Shangyue (Moon Gazing) Building
⑥ 牡丹厅 Peony Hall
⑦ 月亭 Yueting (Moon Pavilion)

扬州何园平面图
Site plan of He Garden, Yangzhou

N

0   10   20m

扬州何园 1-1 剖面图
Section 1-1 of He Garden, Yangzhou

0 5 10m

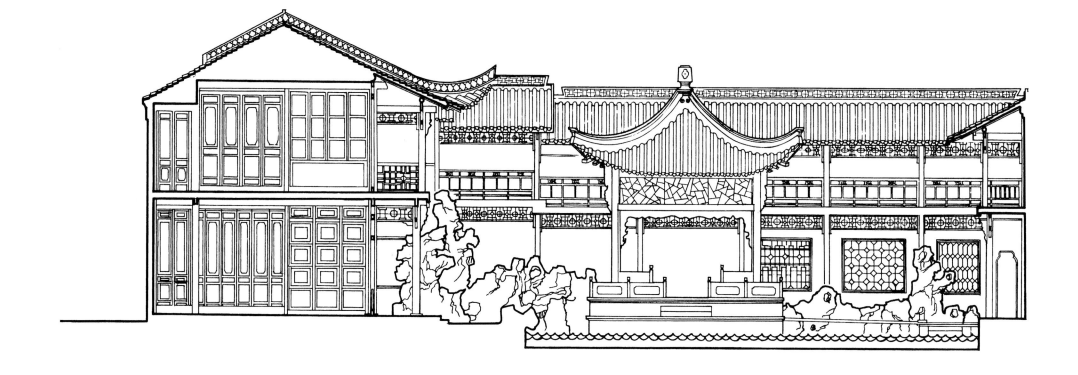

扬州何园 2-2 剖面图
Section 2-2 of He Garden, Yangzhou

0　　　　　5　　　　　10m

扬州何园 3-3 剖面图
Section 3-3 of He Garden, Yangzhou

0 5 10m

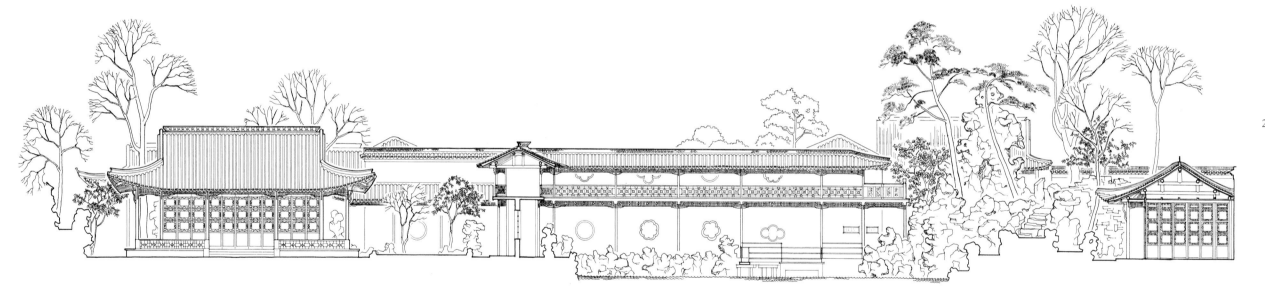

扬州何园 4-4 剖面图
Section 4-4 of He Garden, Yangzhou

0　　5　　10m

静香轩与月亭
Jingxiangxuan and Yueting

Jingxiangxuan (Serene Scent Pavilion) is the central building of He Garden's eastern section. The four-sided hall is also dubbed Chuan Ting, or boat hall, alluded by the wave-patterned pavement in front. Behind the hall, tall lake rockery hills rise around its periphery. Accessible by hidden steps, the hilltop sits a round-planned pavilion named Yueting (Moon Pavilion) supported on six columns.

静香轩是何园东区的主体建筑，位置居中，四面厅式样。厅前地面铺作波浪形图案，又称『船厅』。厅后叠湖石假山，有磴道上下。登山而东，至一六柱圆亭，名『月亭』。

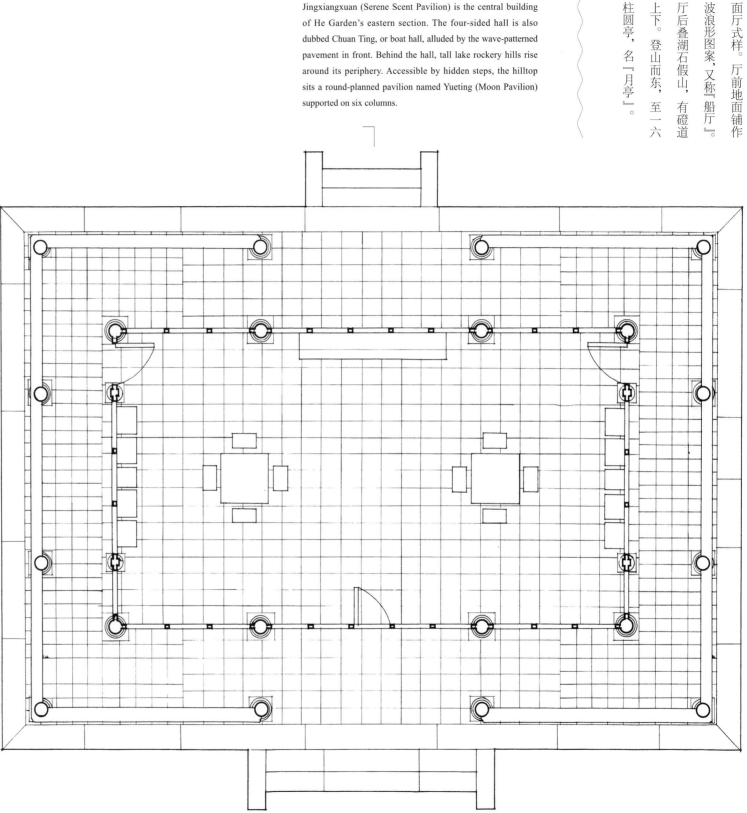

静香轩平面图
Site plan of Jingxiangxuan

N

0　1　2m

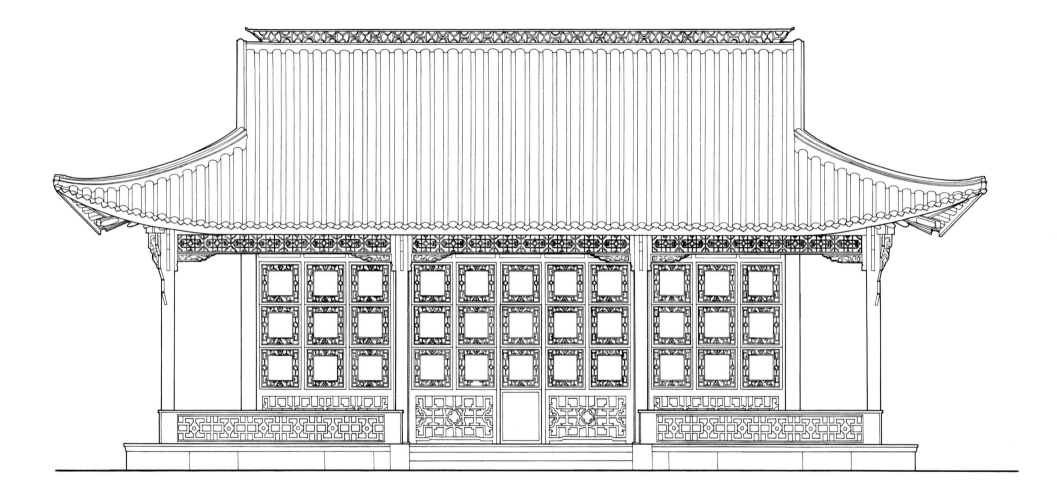

静香轩南立面图
Elevation of Jingxiangxuan

0　　1　　2m

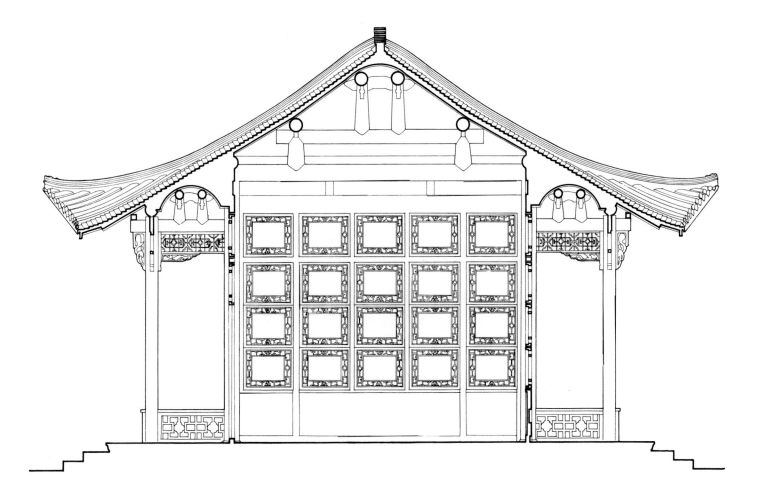

静香轩剖面图
Section of Jingxiangxuan

0 1 2m

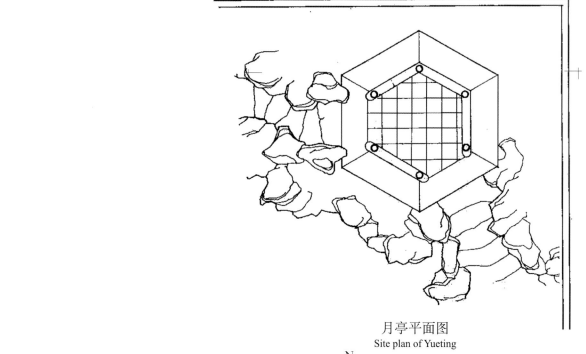

月亭平面图
Site plan of Yueting

N

0　1　2　3m

月亭立面图
Elevation of Yueting

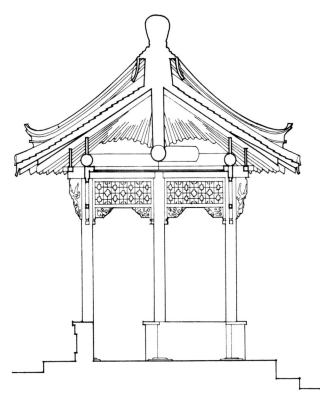

月亭剖面图
Section of Yueting

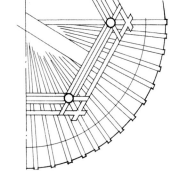

月亭仰视平面图
Reflected ceiling plan of Yueting

水心亭
Shuixin Pavilion

Shuixin (Heart of Water) Pavilion sits in He Garden's western section. Visible from the two-storied, double-sided gallery on the periphery, the large, square-planned pavilion sits openly on the east of the pond. To the pond's west lies a rockery hill stretching southward; to the pond's north is Hudie (Butterfly) Building connected to the two-storied gallery. This gallery runs southward, turning west before turning south again, terminating at Shangyue (Moon Gazing) Building in the southwest corner. Wrapped around the building, the gallery serves as its corridor accessed from both levels.

水心亭位于何园西部片区。由复廊之门转入西部，中央蓄水池，池面较开阔；池东水中设一方亭，体量较大，称「水心亭」。池西叠湖石假山一座，山体向南延伸；池北一楼，名为「蝴蝶厅」，楼与二层的廊连；池南为复廊之延伸，折而向西又向南，楼上楼下均可与西南隅的赏月楼相通。

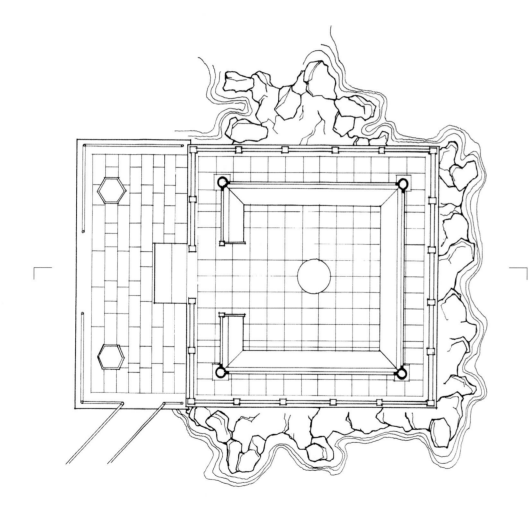

水心亭平面图
Site plan of Shuixin Pavilion

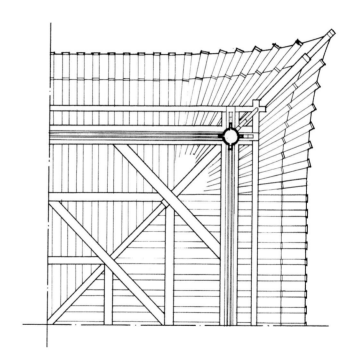

水心亭局部仰视平面图
Reflected ceiling plan of Shuixin Pavilion

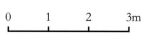

N

0　1　2　3m

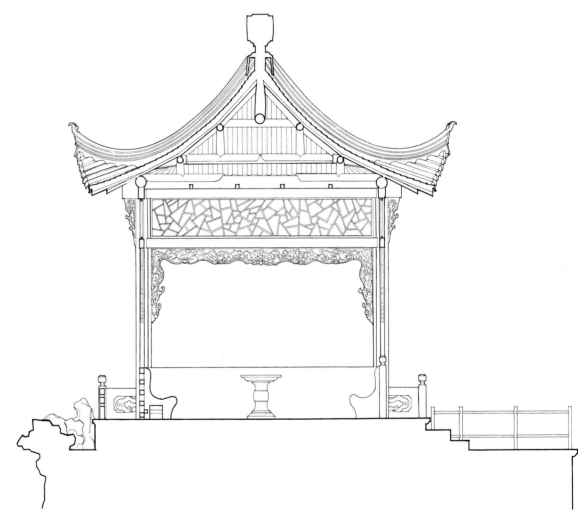

水心亭立面图
Elevation of Shuixin Pavilion

水心亭剖面图
Section of Shuixin Pavilion

Ge Garden, Yangzhou

Ge Garden, located on Dongguan Street of Yangzhou, was built by a salt merchant named HUANG Zhiyun from 1796 to 1820. It was originally the site of Shouzhi (Ancient Angelica Root) Garden with a rockery hill built by SHI Tao in early Qing dynasty. Overtime, the decline in Yangzhou's salt trade resulted in frequent changes of the garden's ownerships, leading it to deteriorate over time. It was finally taken over by landscape authorities in 1979 for restoration and later opened to the public. It is currently listed as a National Priority Protected Site.

Covering an area of a hectare, the garden is known for its seasonal scenes of rockery hills. Entered from the south, one is greeted by flower beds of bamboo and stalagmites, depicting a spring mountain scene. Walking inward, one arrives at a flower hall named Yiyuxuan (Fine Rain Pavilion). Across its northwest pond sits a rockery hill; its bulging form imitates that of a rain cloud's, lending its name *xiashan* (summer mountains). In the northeast of the hall, a massive rockery range faces west; at dusk, sunlight bathes it in a warm autumn hue, painting an image of autumn mountains. Sitting east of Yiyvxuan is Toufeng louyuexuan (Whistling Wind and Glimpse of Moon Pavilion). At three rooms wide, the pavilion's high wall lean against a small rockery hill stacked with snow-white quartzite, evoking the sparkling gleam of fresh winter snow. The twenty-four holes on the wall allows winds to pierce through, its chilling screech reminiscent of those amidst winter mountains. Out of the four mountains, the summer and autumn ones are considerably larger than their spring and winter counterparts.

扬州个园

个园在扬州城内东关街，为清嘉庆年间（1796—1820年）盐商黄至筠所建。其地原有寿芝园，叠石相传为清初石涛所作。扬州盐业中落后，个园屡易其主，凋零日甚。1979年交园林部门管理，进行全面修整后对外开放，现已列为全国重点文物保护单位。

个园面积一公顷余，以四季假山闻名于世。园门向南，门前两旁花坛上栽竹成林，并植石笋数枚，象征春山。入门为园中花厅『宜雨轩』。厅西北方向隔池有一座湖石假山，湖石形如夏云，石峰多以『云』为名，象征夏山。厅之东北方向叠黄石大假山一座，夕阳映照下犹如秋色满山，象征秋山。厅东侧有屋三间，名『透风漏月轩』，屋后倚高墙以宣石叠小山，宣石色白如雪，且含大量石英颗粒，阳光下闪闪如雪后初晴，墙上凿圆孔24个，风起有声，如寒风呼啸，肃杀之气逼人，象征冬山。四山之中，春山、冬山体量极小，主体乃是夏山与秋山。

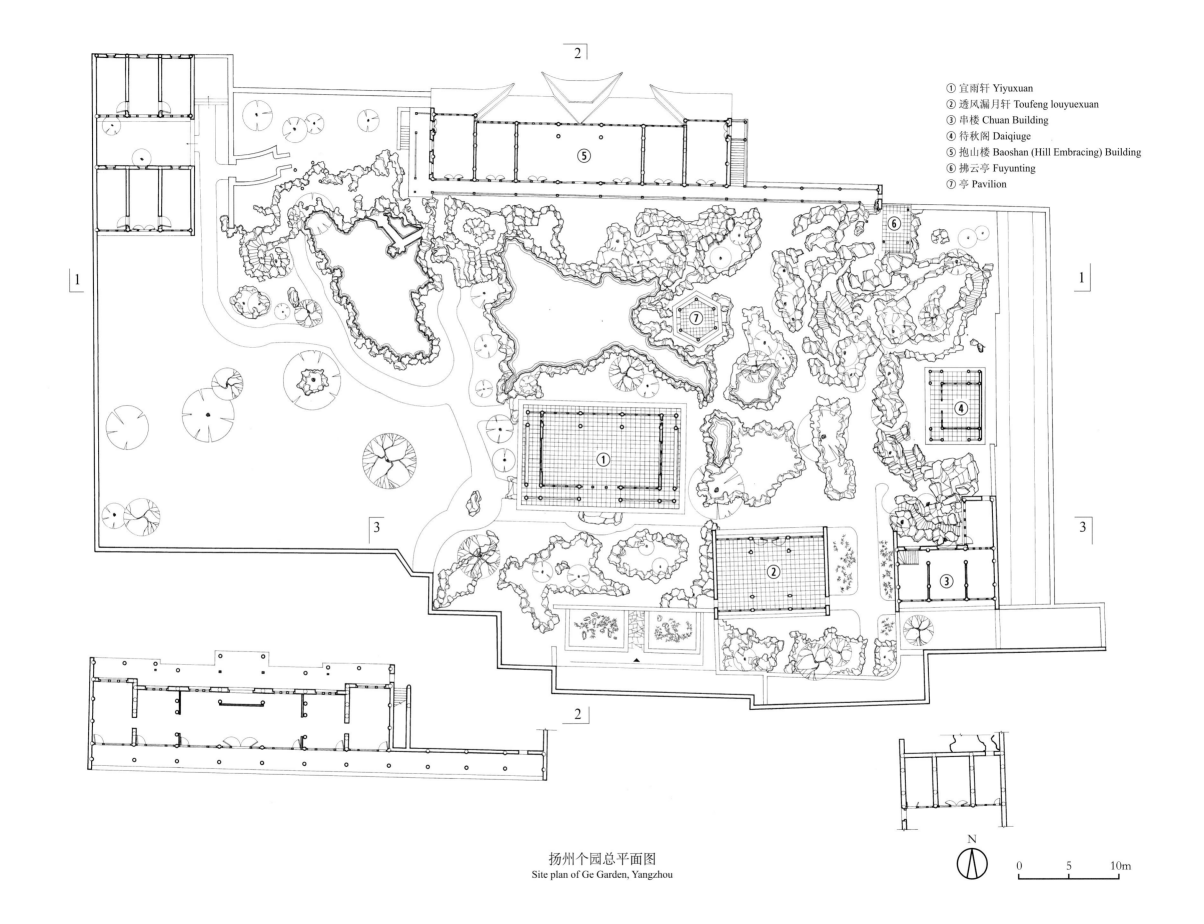

① 宜雨轩 Yiyuxuan
② 透风漏月轩 Toufeng louyuexuan
③ 串楼 Chuan Building
④ 待秋阁 Daiqiuge
⑤ 抱山楼 Baoshan (Hill Embracing) Building
⑥ 拂云亭 Fuyunting
⑦ 亭 Pavilion

扬州个园总平面图
Site plan of Ge Garden, Yangzhou

N

0   5   10m

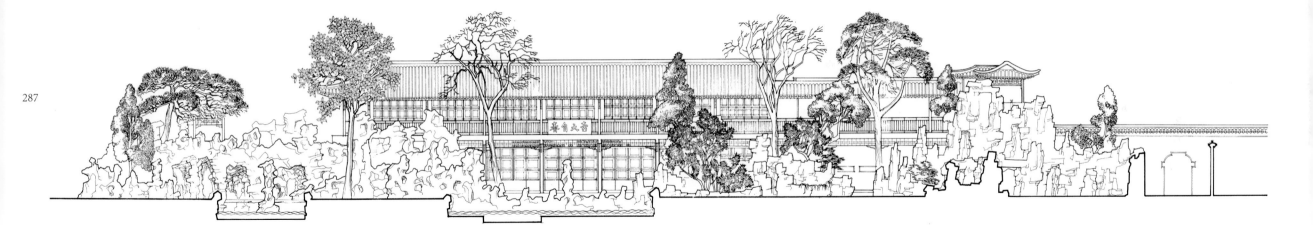

扬州个园 1-1 剖面图
Section 1-1 of Ge Garden, Yangzhou

N

0 1 2 3m

扬州个园 2-2 剖面图
Section 2-2 of Ge Garden, Yangzhou

N

0 1 2 3m

扬州个园 3-3 剖面图
Section 3-3 of Ge Garden, Yangzhou

0 1 2 3m

Yiyuxuan (Fine Rain Pavilion), is a rectilinear flower hall wrapped around by corridors. Originally, the pavilion was to have ground-to-ceiling openings on all sides; however, its incorporation of the rear corridor into the space resulted in some changes. In an attempt to increase depth, the hall extended its partitions northward unto the corridor columns, which led to changing the side walls into row windows and installing *hejingyi* (curved-out railing) along the walkway. Consequently, the design defies the typology of a four-sided pavilion, becoming a variation. In front of the pavilion rests a flower bed made of lake rockeries; it is planted with fragrant sweet olive trees, which perfumes the courtyard upon blossoming every autumn. The southwest corner of the pavilion was once left empty; now several small buildings are erected along a new gallery.

花厅『宜雨轩』采用四面厅式构架，即四面均有走廊，本可四面装落地长窗以利观赏景色，但此厅却将后面走廊围入室内，用以增加厅的进深，并在金柱（步柱）缝上安落地罩分隔空间，东西两侧也改用槛窗，廊内安鹤颈椅，故已失去四面厅之原意，可称之为四面厅之变体。厅前湖石花坛遍植桂花，金秋满园飘香。厅之西南隔原甚空旷，现新增『觅句廊』小楼数间，弥补当年的不足。

宜雨轩平面图
Site plan of Yiyuxuan

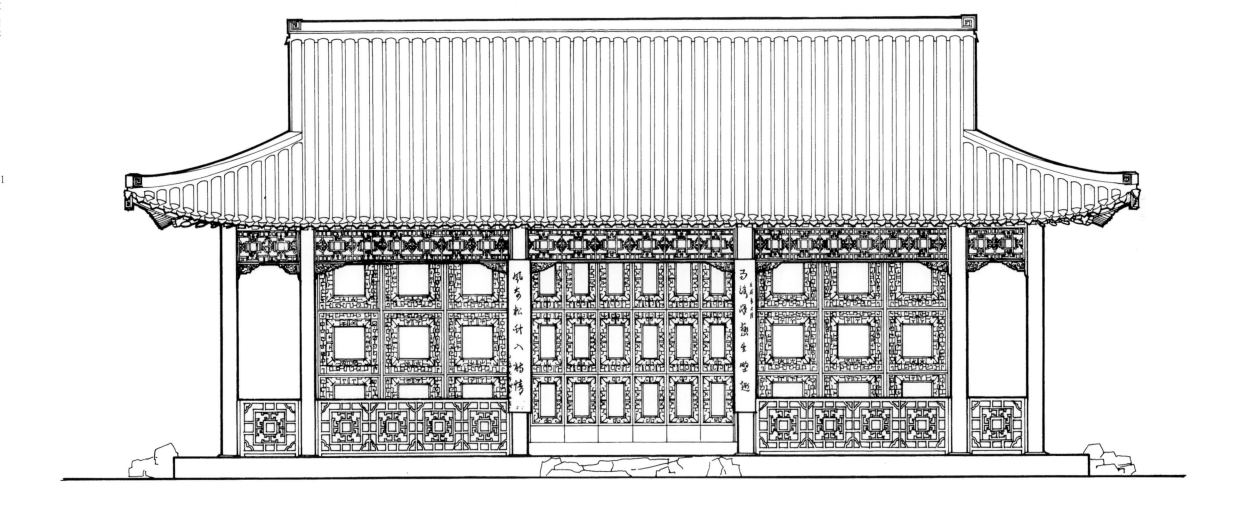

宜雨轩立面图
Elevation of Yiyuxuan

0   1   2   3m

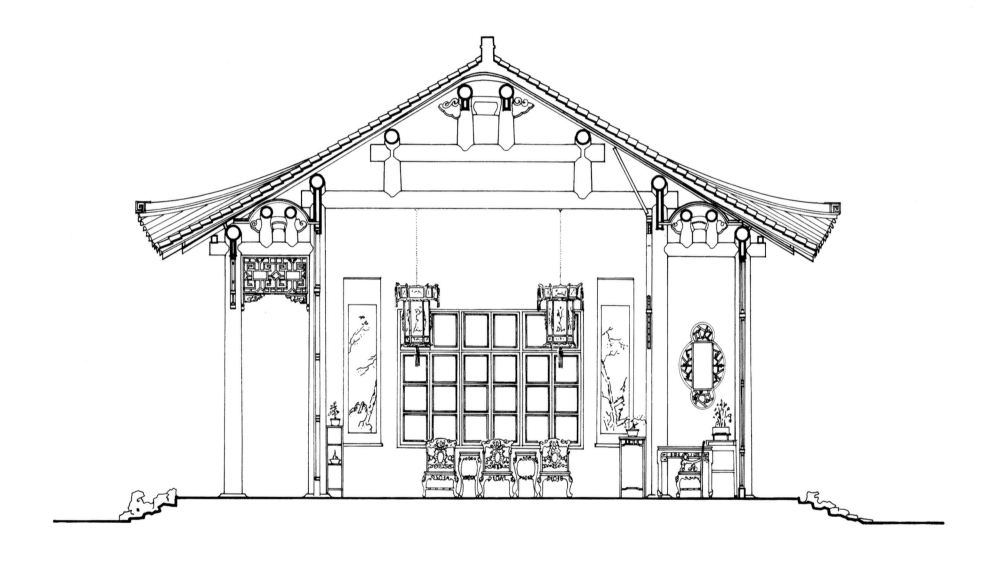

宜雨轩剖面图
Section of Yiyuxuan

0　　1　　2　　3m

拂云亭
Fuyunting

Sitting on top of Ge Garden's autumn mountain, Fuyunting (Sweeping Cloud Pavilion) is connected to the two-storied gallery extended eastward from Baoshan (Hill Embracing) Building. It is the best spot for viewing the garden in its entirety.

拂云亭位于秋山山顶，与抱山楼东端的一段廊楼相连，是全园观景佳处。

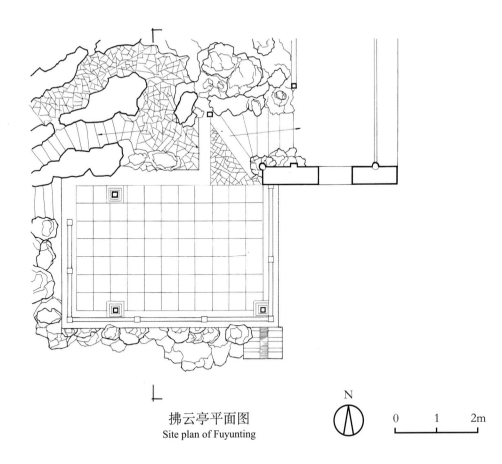

拂云亭平面图
Site plan of Fuyunting

N

0   1   2m

拂云亭立面图
Elevation of Fuyunting

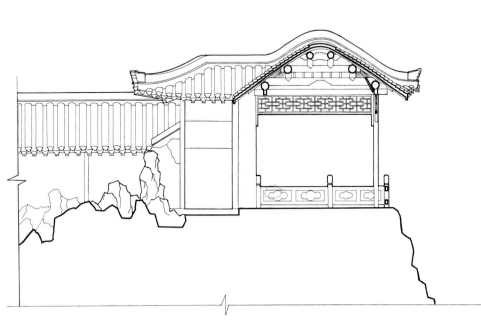

拂云亭剖面图
Section of Fuyunting

Xiaopangu, Yangzhou

Located on Dashu Lane of Yangzhou, Xiaopangu (Small Valley Garden) was built in 1904 by a governor named ZHOU Fu on the site of an older, abandoned garden.

Sitting east from its residence, the narrow garden covers 1,300 m sq within a long and confined space. Inside, rockery hills are pressed against the eastern wall while pavilions and halls sit against the west, separated by a pond in between. Due to the narrow terrain, the flower hall's plan was bent to fit the site. Furthermore, it is built with a *xieshan* (hip-gable) roof on the waterside which turned into a *yingshan* (gable) roof near the wall; its height offsets lends it a distinct visual identity. Its three sides are wrapped by a gallery which continues northward, passes a waterside pavilion and terminates at the small corner building. Turning east, a three-fold zig-zag bridge sits above the pond, leading into caves within the rockeries across. Above the rockery hill is a hexagon pavilion with a panoramic view of the garden. It is reached by stone steps from north of the hill and south from the east gallery. From across the pond, the sight of undulating rockeries appears both real and intangible. A central hill towers above the rockery range, accentuating the subliminal presence of the hill.

扬州小盘谷

小盘谷在扬州大树巷，建于光绪三十年（1904年），为两江总督周馥依旧园重建。

园位于住宅东侧，面积约1300平方米，成南北长、东西窄的长条形，假山靠东墙，亭阁斋馆靠西墙，水池居于两者之间。由于地形狭窄，主体建筑花厅作曲尺形平面，临池一面做歇山顶，而靠墙一面做硬山顶，屋顶错落有致，富于变化。三面有廊，向北延伸，经水榭达北端小楼。水池上架一三折石桥，渡桥可入假山洞。山上平台建六角亭一座，可俯览全园。亭之南北两端有磴道可供上下：南端下至东廊内，北端下至假山北侧。由水池西岸观赏，可见山体起伏，有虚实，中间一峰耸然峙起，颇有雄奇峭拔之势。

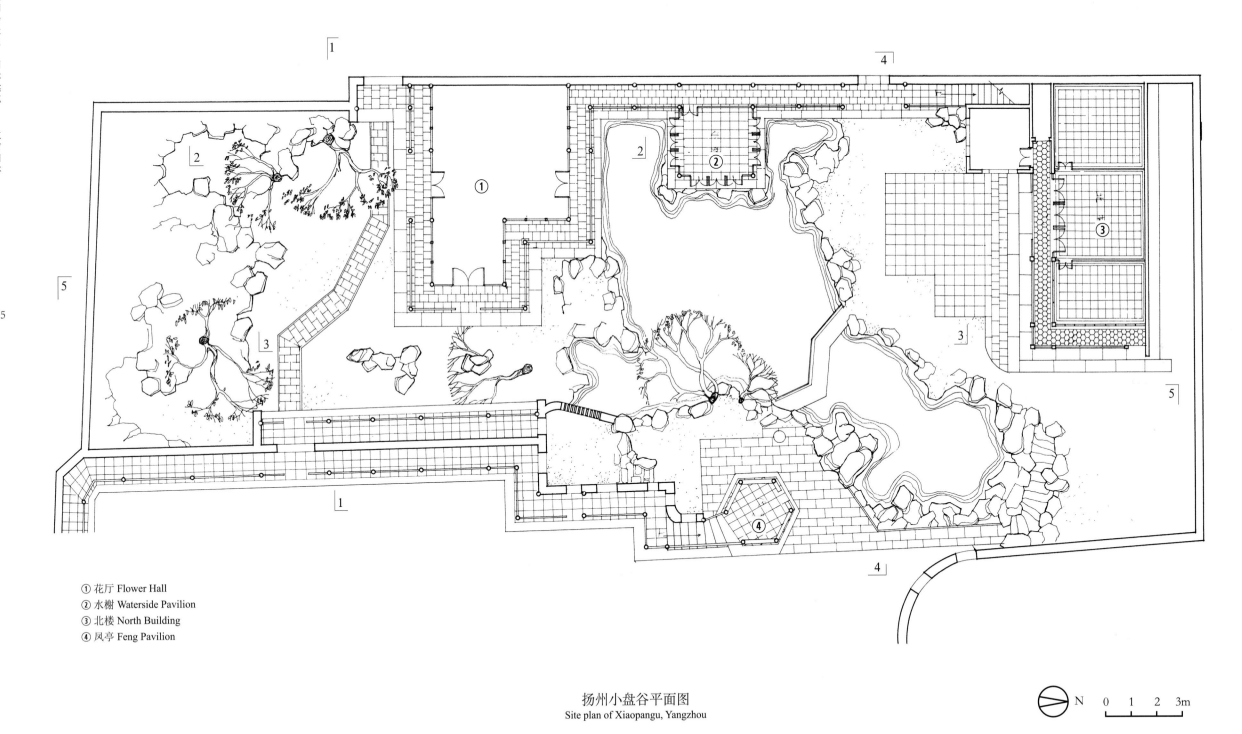

① 花厅 Flower Hall
② 水榭 Waterside Pavilion
③ 北楼 North Building
④ 凤亭 Feng Pavilion

扬州小盘谷平面图
Site plan of Xiaopangu, Yangzhou

N   0 1 2 3m

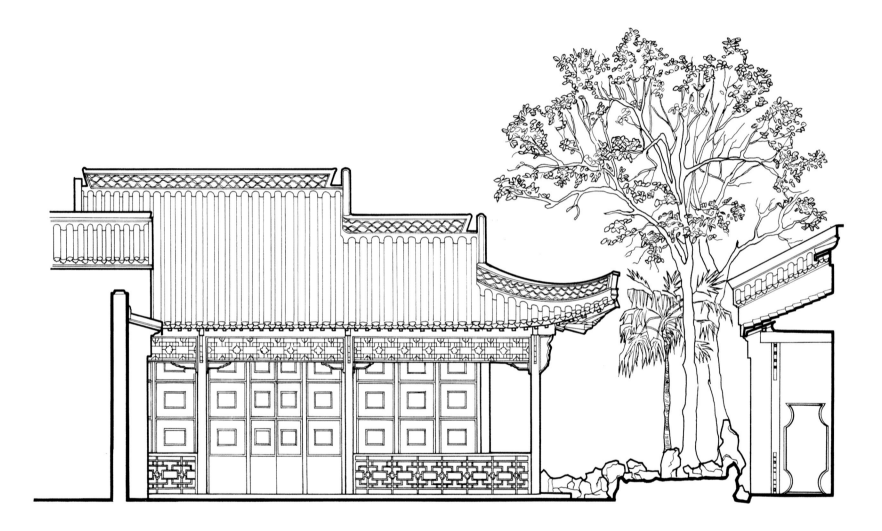

扬州小盘谷 1-1 剖面图
Section 1-1 of Xiaopangu, Yangzhou

0　　1　　2　　3m

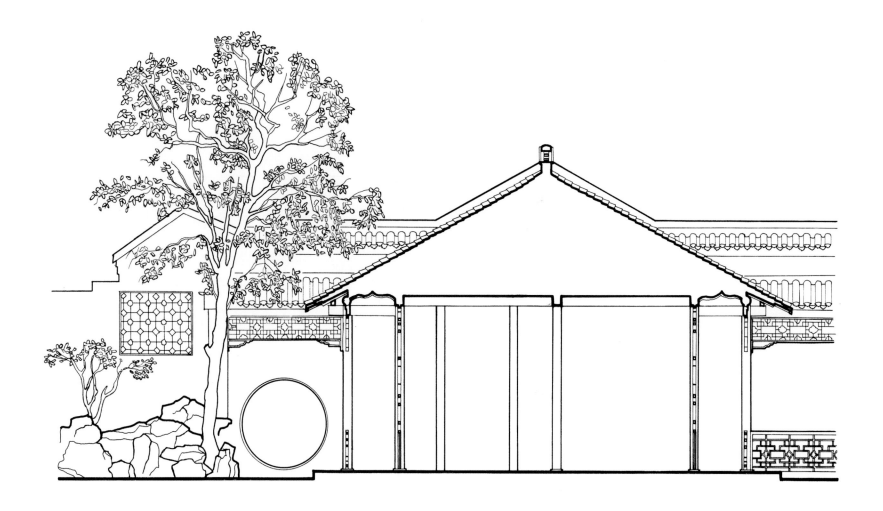

扬州小盘谷 2-2 剖面图
Section 2-2 of Xiaopangu, Yangzhou

0　　　1　　　2　　　3m

扬州小盘谷 3-3 剖面图
Section 3-3 of Xiaopangu, Yangzhou

0　　1　　2　　3m

扬州小盘谷 4-4 剖面图
Section 4-4 of Xiaopangu, Yangzhou

0　　1　　2　　3m

扬州小盘谷 5-5 剖面图
Section 5-5 of Xiaopangu, Yangzhou

0　1　2　3m

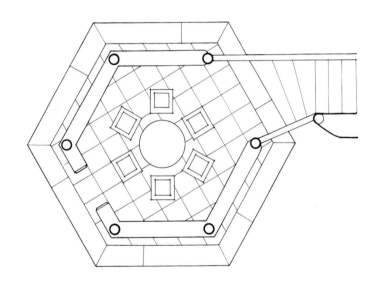

六角亭平面图
Site plan of Feng Pavilion

六角亭立面图
Elevation of Feng Pavilion

Pingshantang West Garden, Yangzhou

Pingshantang (Gentle Mountain Hall) sits west within Daminsi, located on a hill in northwest Yangzhou. It was first built in 1048 by a governor of Yangzhou at the time named OUYANG Xiu. The garden was destroyed in Southern Song dynasty and completely deserted throughout Yuan and Ming dynasties. It was rebuilt in 1673 with the addition of shrine in memory of OUYANG. The buildings today, including Pingshan Hall, Gulin (Ancient Forest) Hall and OUYANG's shrine were all reconstructed between 1862 and 1874. Xiyuan (West Garden), directly west of Pingshan Hall, was built earlier in 1736; it is also named Fangpu (Fragrant Garden).

Covering an area of 1.7 ha, Xiyuan is designed with a limited number of buildings and pavilions. An expansive pond opens in the middle surrounded by bamboo groves and trees, painting a wild forest scene. On an islet of the pond sits a three-room deep pavilion resembling a boat. To the pond's northwest stands Nanmu Hall, its structure originated from a Guanyin'an south of the city. To the pond's south lies Tingshi shanfang (Stone Sound Cottage), a hall built in cypress and moved from the city's old district. Encircling the pond, a continuous pathway links the garden's sceneries altogether. Aside from Diwu quanjing (Fifth Spring) Pavilion on the islet, there are also Kangxi Monument Pavilion, Qianlong Monument Pavilion and Daiyue (Wait for Moon) Pavilion along the pond's east.

扬州平山堂西园

平山堂在扬州西北部蜀岗之上、大明寺西侧，是北宋庆历八年（1048年）欧阳修守扬州时所创。南宋时被毁，元明时湮没无闻，清康熙十二年（1673年）得以重建，并于堂后建欧阳修祠以祀之。今存建筑有平山堂、谷林堂及欧阳修祠，均为同治年间（1862—1874年）重建。西园又名『芳圃』，创于清乾隆元年（1736年）。

现西园面积约1.7公顷，建筑物较少，疏布亭阁。水池居中，水面开阔，竹树茂密，富山林野趣。池中有一岛，岛上作船厅三间。池西北筑北厅，是从城南观音庵移来的楠木厅。池南听石山房是从旧城移来的柏木厅。环池辟山间游路一周，使游人能穿林越岗，遍历园中胜境。园内除池中第五泉井亭外，还有东岸的康熙碑亭、乾隆碑亭和待月亭各一座。

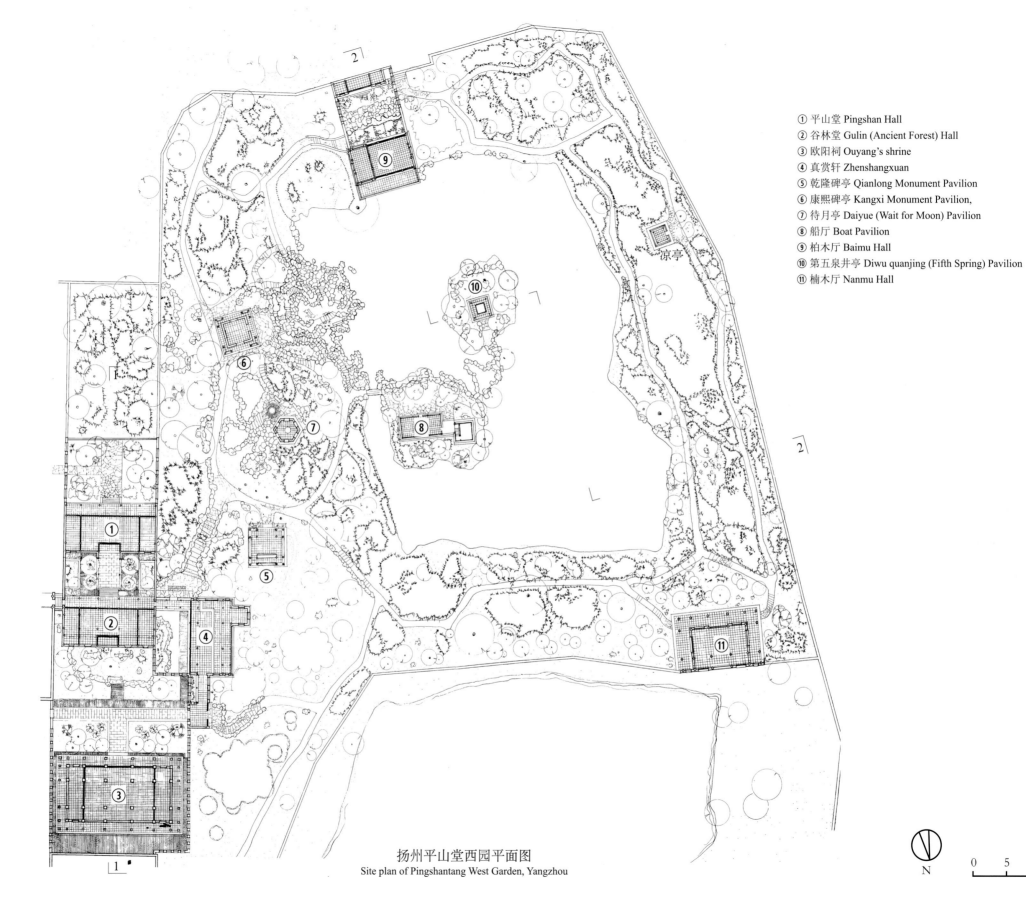

① 平山堂 Pingshan Hall
② 谷林堂 Gulin (Ancient Forest) Hall
③ 欧阳祠 Ouyang's shrine
④ 真赏轩 Zhenshangxuan
⑤ 乾隆碑亭 Qianlong Monument Pavilion
⑥ 康熙碑亭 Kangxi Monument Pavilion,
⑦ 待月亭 Daiyue (Wait for Moon) Pavilion
⑧ 船厅 Boat Pavilion
⑨ 柏木厅 Baimu Hall
⑩ 第五泉井亭 Diwu quanjing (Fifth Spring) Pavilion
⑪ 楠木厅 Nanmu Hall

凉亭

扬州平山堂西园平面图
Site plan of Pingshantang West Garden, Yangzhou

N

0   5   10m

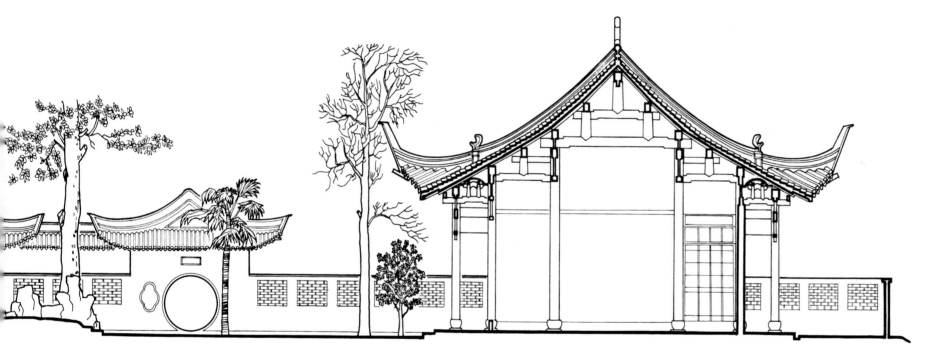

0　1　2　3m

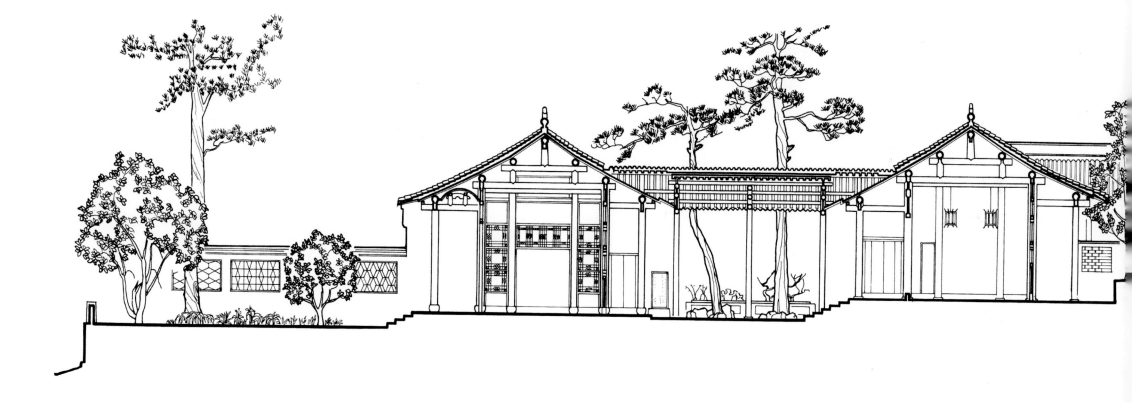

扬州平山堂西园 1-1 剖面图
Section 1-1 of Pingshantang West Garden, Yangzhou

0　1　2　3m

扬州平山堂西园 2—2 剖面图
Section 2-2 of Pingshantang West Garden, Yangzhou

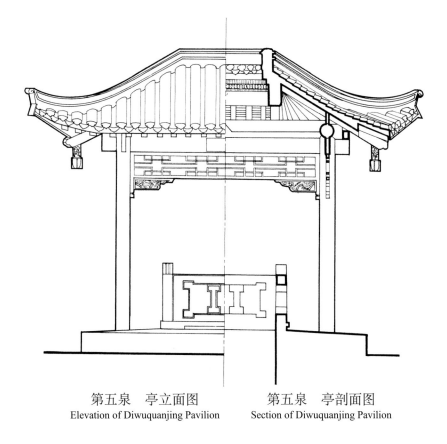

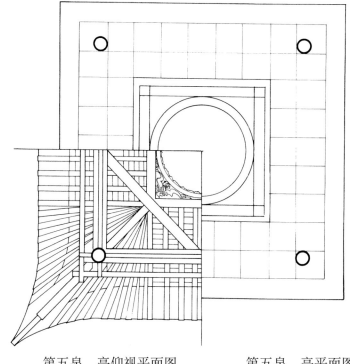

第五泉　亭立面图
Elevation of Diwuquanjing Pavilion

第五泉　亭剖面图
Section of Diwuquanjing Pavilion

第五泉　亭仰视平面图
Reflected ceiling plan of Diwuquanjing Pavilion

第五泉　亭平面图
Site plan of Diwuquanjing Pavilion

N

0　　　　1　　　　2m

Qiao Garden, Taizhou

Located on Hailing South Road of Taizhou, the garden was originally named Rishe (Daily Stroll) Garden, built by a scholar named CHEN Yingfang in 1574. After frequent changes in ownership, it was renamed Sanfengyuan (Three Peaks Garden) during the reign of Qing emperor Jiaqing. Later, it was reclaimed by salt transport supervisor (official title: an *yanyushi*) named QIAO Songning, thus renamed Qiao Garden.

The garden sits behind the now-demolished residence; covering 1,300 m sq in area, it is divided into northern and southern sections. The larger, southern section is marked by the flower hall Shanxiang caotang (Mountain Sound Thatch Hall). In front of the hall lies a pond and a stacked rockery hill across from it, both serving as the hall's facing scene. On the hilltop stands an ancient cypress as old as the garden, dating the hill's origin to the reign of Ming emperor Wanli. The hill also has two pavilions: Shuyuting (Counting Fish) Pavilion to its east and Banfang (Half-square) Pavilion to its west. Other elements on the hill include stone footbridges, three stalagmite pillars, and a spring well. The northern section, though smaller than the southern one, is situated on a raised terrain, lending the garden an ascendance in views as one move northward. The northern section's main building is Gengji Hall, flanked by Songchui Building on the east and Yinchao Pavilion on the west. Brick-mortar wall, perforated roof ridge and perforated brick openings were applied uniformly throughout the garden's architecture.

泰州乔园

位于泰州市区海陵南路，由明万历二年（1574年）进士陈应芳创建，题名『日涉园』。其后屡易其主，清嘉庆年间（1796—1820年）改『三峰园』，后又归两淮盐运使乔松年，俗称『乔园』。

园在住宅后部，现住宅已大部分拆除。全园面积约1300平方米，分南北两区。南区较大，以花厅『山响草堂』为主体，厅前凿池，堆湖石假山，构成花厅对景。山上留有古柏一株，其树龄与园龄相当，可证此山仍为万历旧物。山上东西各有一亭，东为数鱼亭，西为半方亭。山上有石梁，有石笋三株，还有一井泉眼。北区较小而基址增高，故全园呈前卑后昂之势。北区以居中的『绠汲堂』为主体，左设小楼『松吹阁』，右建小斋『因巢亭』。园中建筑物均用清水砖墙、空花脊及青砖漏窗。

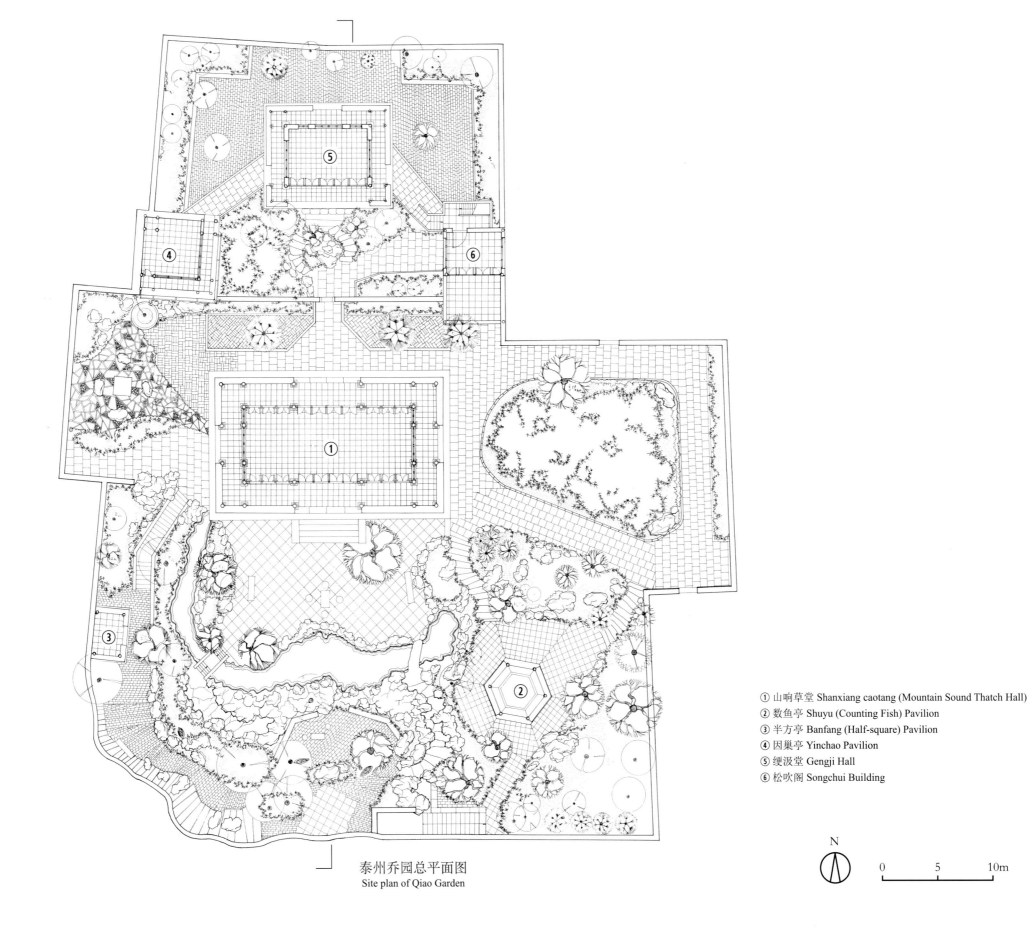

① 山响草堂 Shanxiang caotang (Mountain Sound Thatch Hall)
② 数鱼亭 Shuyu (Counting Fish) Pavilion
③ 半方亭 Banfang (Half-square) Pavilion
④ 因巢亭 Yinchao Pavilion
⑤ 绠汲堂 Gengji Hall
⑥ 松吹阁 Songchui Building

泰州乔园总平面图
Site plan of Qiao Garden

N

0   5   10m

泰州乔园剖面图
Section of Qiao Garden

0 2 4m

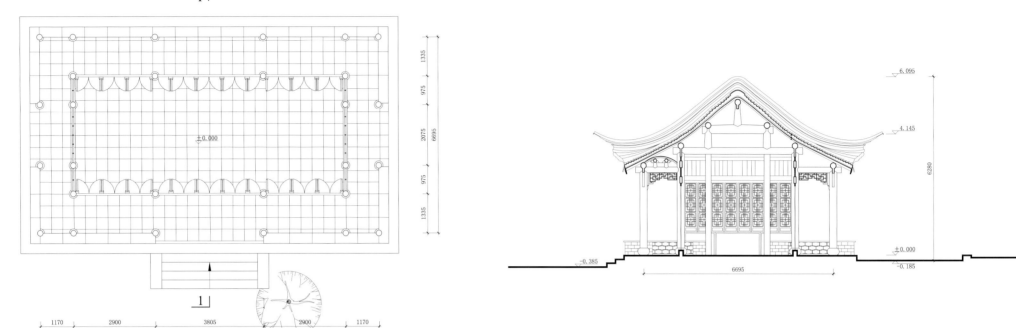

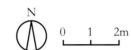

山响草堂平面图
Site plan of Shanxiang caotang

山响草堂 1-1 剖面图
Section 1-1 of Shanxiang caotang

N

0　1　2m

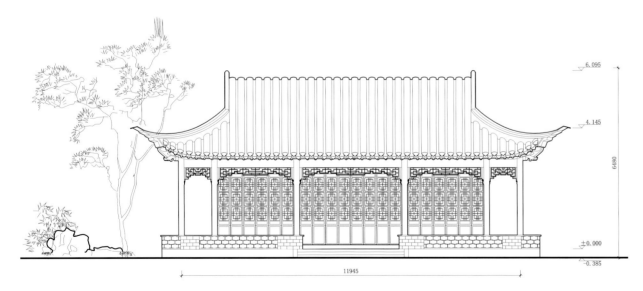

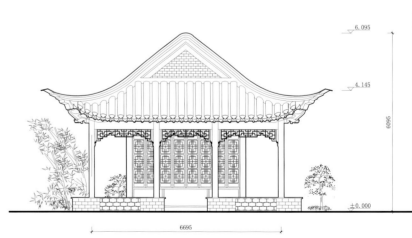

山响草堂南立面图
South elevation of Shanxiang caotang

山响草堂西立面图
West elevation of Shanxiang caotang

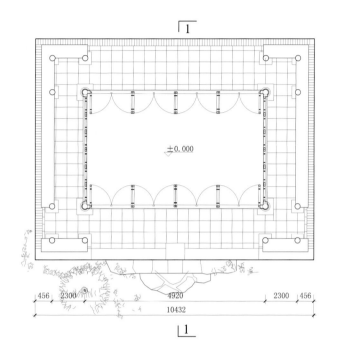

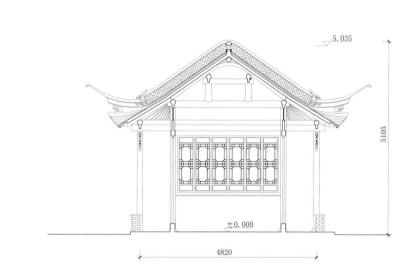

绠汲堂平面图
Site plan of Gengji Hall

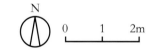

绠汲堂 1-1 剖面图
Section 1-1 of Gengji Hall

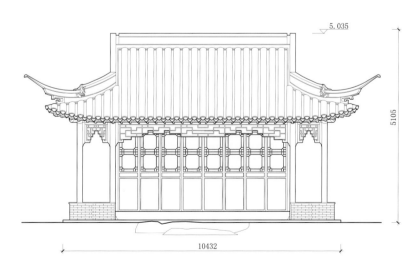

绠汲堂南立面图
South elevation of Gengji Hall

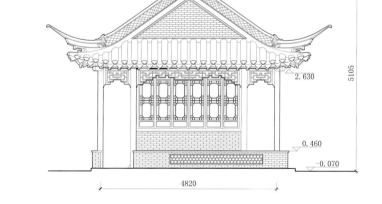

绠汲堂东立面图
East elevation of Gengji Hall

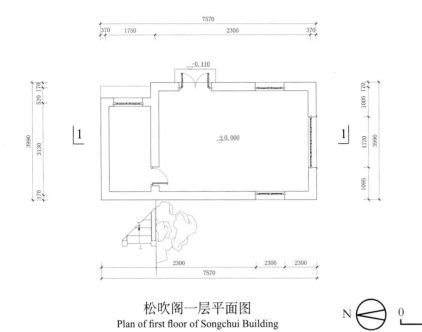

松吹阁一层平面图
Plan of first floor of Songchui Building

松吹阁 1-1 剖面图
Section 1-1 of Songchui Building

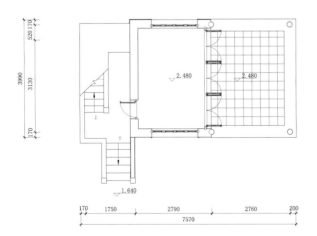

松吹阁二层平面图
Plan of second floor of Songchui Building

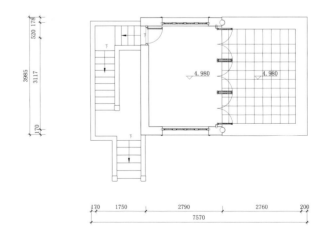

松吹阁三层平面图
Plan of third floor of Songchui Building

中国古建筑测绘大系·园林建筑——江南园林

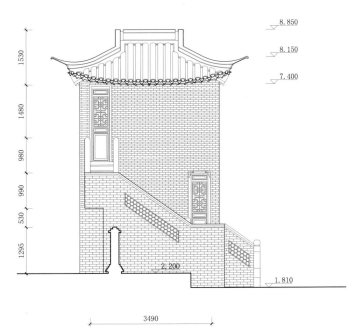

松吹阁北立面图
North elevation of Songchui Building

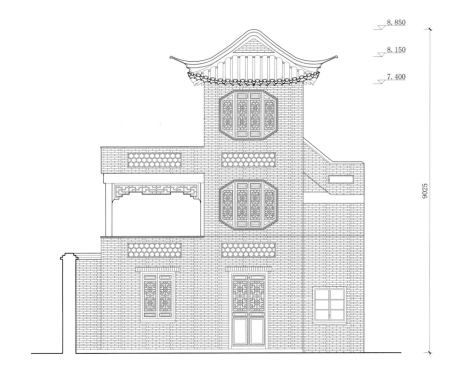

松吹阁东立面图
East elevation of Songchui Building

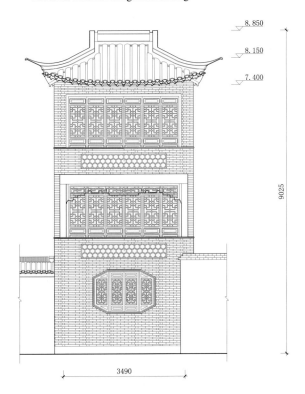

松吹阁南立面图
South elevation of Songchui Building

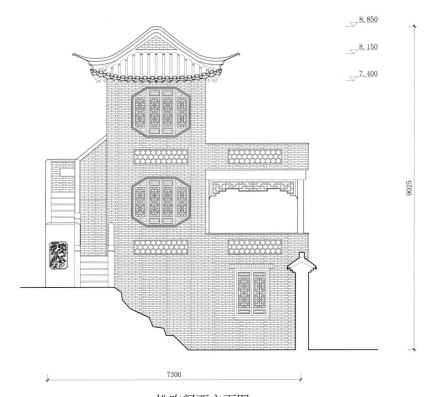

松吹阁西立面图
West elevation of Songchui Building

0　　1　　2m

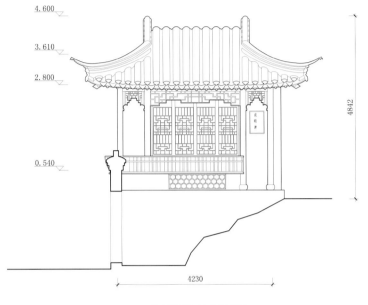

因巢亭东立面图
East elevation of Yinchao Pavilion

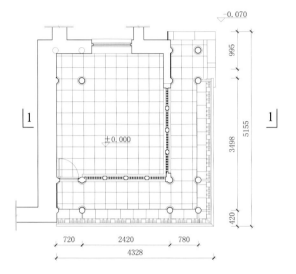

因巢亭平面图
Site plan of Yinchao Pavilion

0   1   2m

N

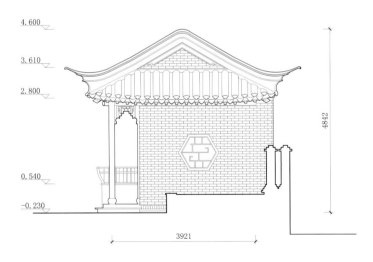

因巢亭北立面图
North elevation of Yinchao Pavilion

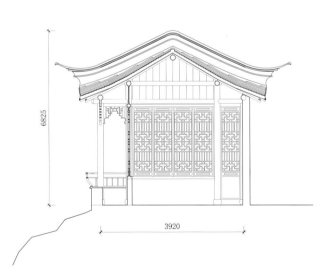

因巢亭 1-1 剖面图
Section 1-1 of Yinchao Pavilion

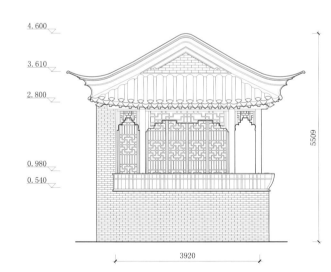

因巢亭南立面图
South elevation of Yinchao Pavilion

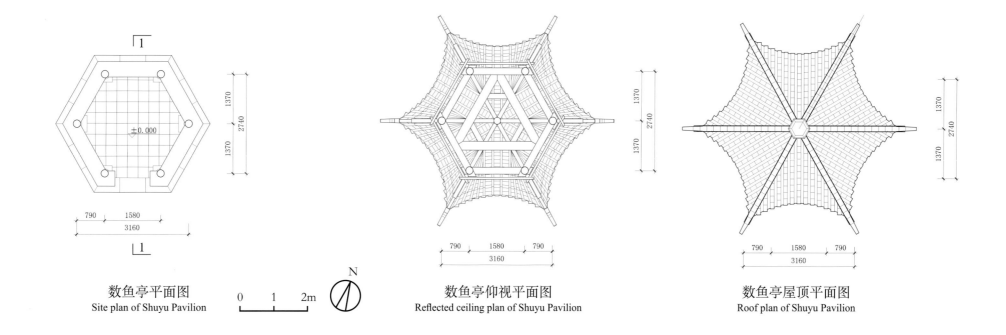

数鱼亭平面图
Site plan of Shuyu Pavilion

0   1   2m

N

数鱼亭仰视平面图
Reflected ceiling plan of Shuyu Pavilion

数鱼亭屋顶平面图
Roof plan of Shuyu Pavilion

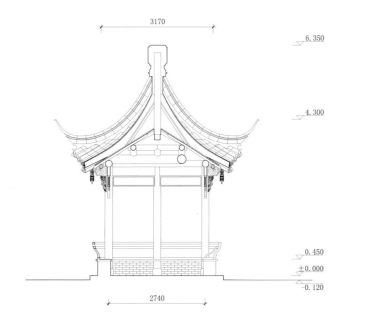

数鱼亭 1-1 剖面图
Section 1-1 of Shuyu Pavilion

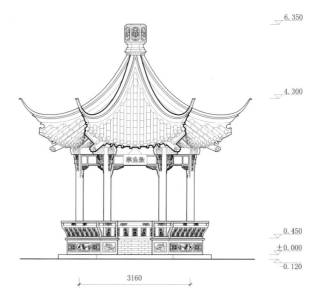

数鱼亭立面图
Elevation of Shuyu Pavilion

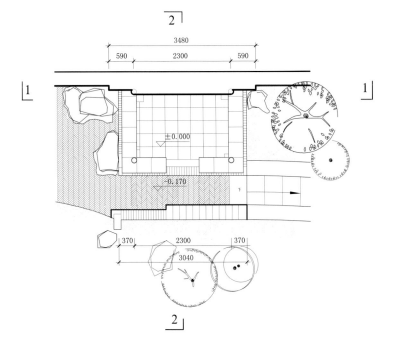

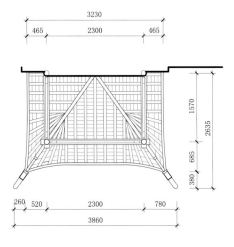

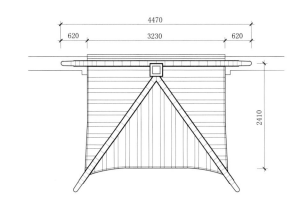

半方亭平面图
Site plan of Banfang Pavilion

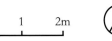

半方亭仰视平面图
Reflected ceiling plan of Banfang Pavilion

半方亭屋顶平面图
Roof plan of Banfang Pavilion

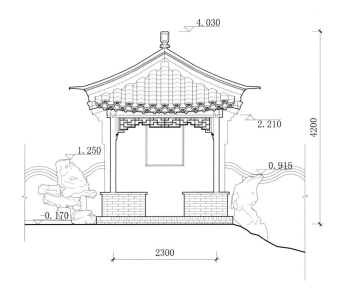

半方亭 1-1 剖面图
Section 1-1 of Banfang Pavilion

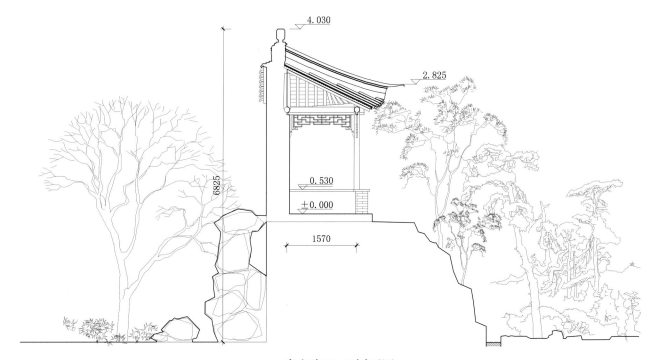

半方亭 2-2 剖面图
Section 2-2 of Banfang Pavilion

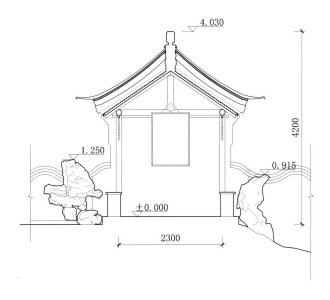

半方亭东立面图
East elevation of Banfang Pavilion

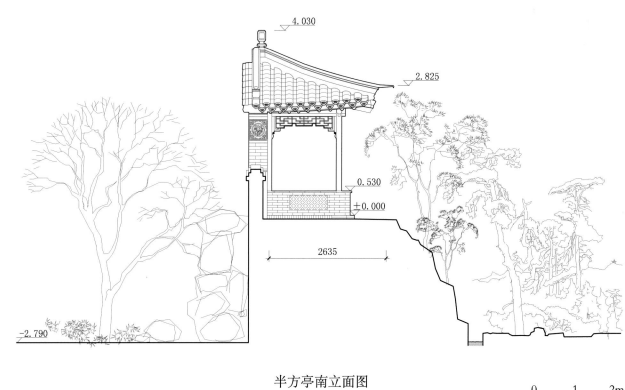

半方亭南立面图
South elevation of Banfang Pavilion

0　1　2m

Qi Garden, Haiyan

Located in Wuyuan Town of Haiyan, Qi (Enchanting) Garden is a well-preserved, late-Qing private garden in Zhejiang Province.

The garden is located north of its residence; covering an area of one hectare, it is 120 m from north to south and 80 m from east to west. The main building, Tanying (Water Shade) Pavilion, is a three-bays wide, four-sided structure sitting in the south. A sparse rockery hill is located to its north and a pond to its south; while the southern yard is narrow, the northern one is more spacious. With an extended stream, the southern pond flows into the larger northern one. In the northern pond, stacked rockery hills runs from north to east then south; its west is traced by a secluded trail leading to a waterside pavilion. Long dikes divide the water into ponds of different sizes. Throughout the garden, the presence of countless ancient trees paints a rare and remarkable sight. Amidst it all, the upper portion of most rockery hills are stacked with lake rockeries. Their enormous scale, undulating form, deep caverns, twisting steps, and illusive offsets created a naturalistic scene with the utmost refinement in gardening techniques. On the northern hill is Yiyun (Still Cloud) Pavilion. Supported on six columns, its steep, hexagonal-pyramidal roof is subtly fused to the subliminal air of the hills.

海盐绮园

绮园在浙江海盐武原镇内，是浙江保存较好的一座清末私家园林。

园在住宅之北，占地约1公顷。南北长120米，东西阔80米。主体建筑『潭影轩』在南部，系三开间周围廊之四面厅。轩北为山，南为池，中南部紧凑，北部宽敞。轩南水池以溪涧形式与北部大池相通。北部水池之南、之北、之东三面叠山，池西为小径，为水榭。北部水面又以长堤将水面再分大小几处。园内大量古树古藤为他处难得，十分珍贵。园中假山，上部皆以湖石堆成，规模宏大，姿态逶迤，洞壑深幽，磴道盘旋，石梁高低变幻，极尽人工仿天然之能事。北山巅上有依云亭，六柱攒尖，屋盖陡峻，与山势一气呵成。

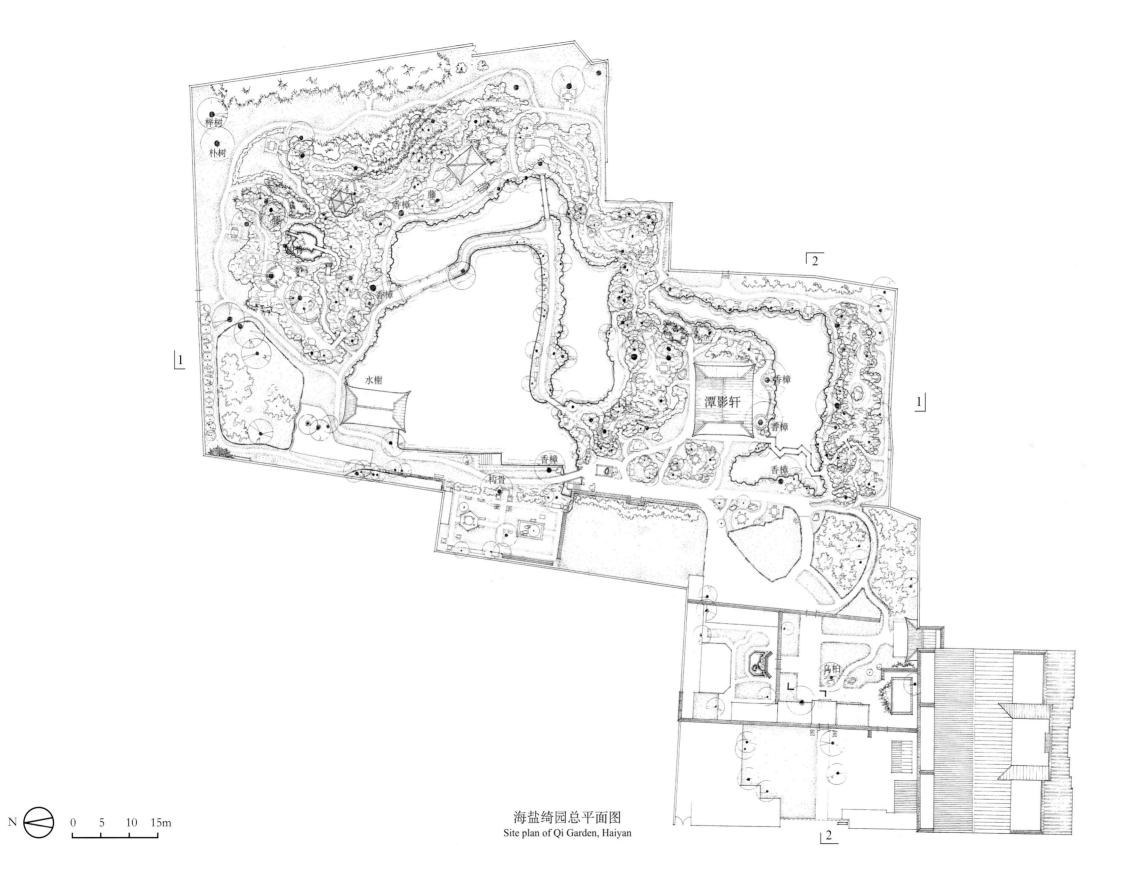

榉树

朴树

香樟

水榭

香樟

潭影轩

香樟

香樟

香樟

香樟

构雪

乌柏

N

0 5 10 15m

海盐绮园总平面图
Site plan of Qi Garden, Haiyan

海盐绮园透视图
Perspective of Qi Garden, Haiyan

海盐绮园 1-1 剖面图
Section 1-1 of Qi Garden, Haiyan

海盐绮园 2-2 剖面图
Section 2-2 of Qi Garden, Haiyan

0　　5　　10m

GUO's Residence, Hangzhou

GUO's Residence, also known as Fenyang Villa, was originally named Song's Residence. It is located on the north of Wolong Bridge along Xishan Road, bordering the western bank of West Lake in Hangzhou. The garden was originally built by SONG Duanfu, a silk merchant from Qing dynasty who named it Duanyou Villa; it was later handed to another owner with the surname GUO, hence GUO's Residence. Later, the garden was severely damaged; although a survey made in the 1980s revealed traces of its foundation, it was left in ruin for some time. It was finally restored and opened to the public in the 1990s. With its original layout intact, interventions were kept to a minimum. As a result, the wooden structures appear austere yet elegant and the garden refined yet timeless.

GUO's Residence is laid out along the lake from north to south. It features lush greenery, secluded trails and a pond flowing westward, fed by an inlet from the south. Upon entering from the north, the view opens unto Yijing tiankai (Mirrored Sky) Pond; after passing Quqiao (Winding Bridge), one arrives at Liangyixuan (Two-sided Pavilion), which divides the water into southern and northern halves. The southern section contains a group of closed courtyards further south; as for the northern section, an open gallery lines the pond. Along the garden's western wall, pavilions and parallel galleries—sometimes detached and sometimes bent—defined the walking route. Along the eastern perimeter, there are three spots through which the water flows to or from the West Lake: beneath the southern arched gallery bridge, a central rockery cave, and the northern waterside platform. Most buildings are arranged along the lakeside: jutting out and retreating, opening then closing. In between, long walls are punctuated by perforated openings and waterside platform. Dotted by stacked rockeries and greenery, the resulting scenes are fragmented yet continuous, brought close then recessed, tangible and ephemeral. An interplay of height is exemplified between Shangxin yuemu (Pleasing View) Pavilion on a rockery hill and a lakeside pavilion. From the embankment, one is able to borrow a glimpse of both the Su Causeway and the ripples from afar.

杭州郭庄

郭庄一名『汾阳别墅』，昔之宋庄也，地处杭州西湖西山路卧龙桥北，滨里西湖之西岸。园原为清代绸商宋端甫所筑，称『端友别墅』后归郭氏。郭庄曾一度毁坏严重，20世纪80年代测绘时虽存旧迹，但仍是一片荒凉。20世纪90年代后经修缮，已对外开放，且保留旧时格局，木构少做油漆，水木清华，雅洁淡泊，富有古趣。

郭庄依南北向沿湖长向布局。园中树木茂盛，小路幽曲，湖水由西向东于南墙外汇集成池。进园后『一镜天开』水池寥廓，经由曲桥，抵跨于水上的『两宜轩』，此轩将大水面分为南北两处。隔南池的南端为一组合院落，相对封闭，院北侧沿池则有空透的廊子。园之西墙高而平直，沿墙或即或离，或直或曲地布局亭廊，形成游线。东边沿湖自南而北有三处将西湖水通至园内：流经桥下、湖石假山洞下和平台下。而沿湖的建筑设置，则进退有致、敞闭自如。有的是设墙开漏窗，有的是平台临水，其间叠以山石，植以树木，断而有续，前后参差，有实有虚。在高低关系上，有建于水洞假山上的『赏心悦目阁』，也有把酒临风的水榭。远可借苏堤，近可瞰湖波。

① 门厅 Entrance Hall
② 一镜天开 Yijing tiankai (Mirrored Sky) Pond
③ 两宜轩 Liangyixuan (Two-sided Pavilion)
④ 雪香分春 Xuexiangfenchun
⑤ 雨山爽气 Yushanshuangqi
⑥ 乘浪起风 Chenglangqifeng
⑦ 舒卷自如 Shujuanziru
⑧ 锦苏楼 Jinsu Building
⑨ 赏心悦目阁 Shangxin yuemu (Pleasing View) Pavilion
⑩ 曲桥 Quqiao (Winding Bridge)
⑪ 西湖 The West Lake

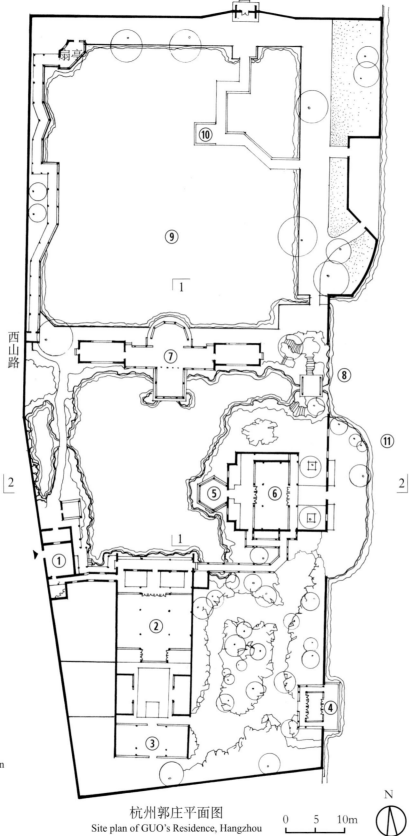

杭州郭庄平面图
Site plan of GUO's Residence, Hangzhou

0   5   10m

N

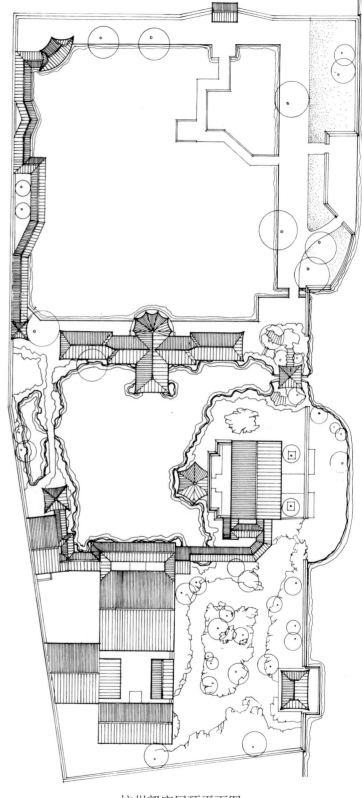

杭州郭庄屋顶平面图
Roof plan of GUO's Residence, Hangzhou

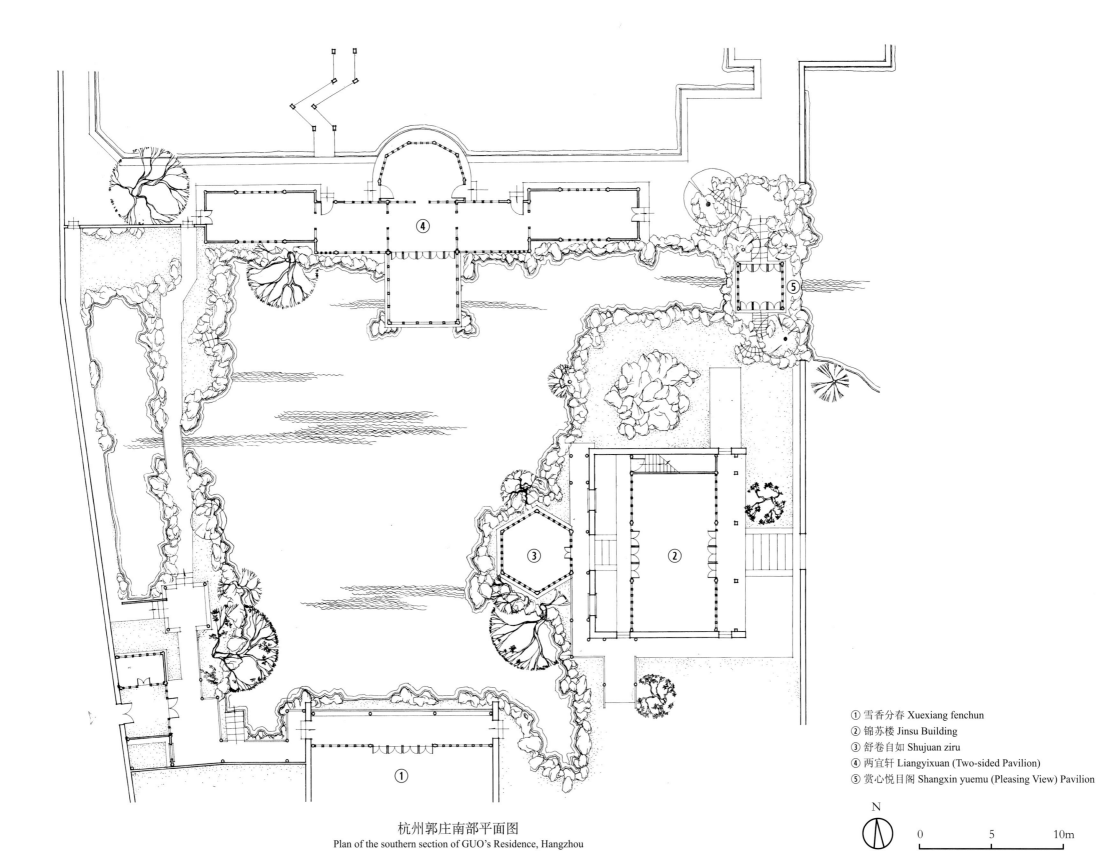

① 雪香分春 Xuexiang fenchun
② 锦苏楼 Jinsu Building
③ 舒卷自如 Shujuan ziru
④ 两宜轩 Liangyixuan (Two-sided Pavilion)
⑤ 赏心悦目阁 Shangxin yuemu (Pleasing View) Pavilion

杭州郭庄南部平面图
Plan of the southern section of GUO's Residence, Hangzhou

N

0   5   10m

杭州郭庄 1-1 剖面图
Section 1-1 of GUO's Residence, Hangzhou

0 1 2 3m

杭州郭庄 2—2 剖面图
Section 2-2 of GUO's Residence, Hangzhou

0　1　2　3m

景园

Landscape Parks

Landscape parks are usually situated around pre-existing hills or water bodies and open to the public. Pavilions, buildings, as well as monasteries and temples would be built amidst the spacious land to create a scenic spot for sightseeing. They are composed as a complex with defined boundaries and easy accesses to the sceneries.

A landscape park is often expansive and located in suburban areas, such as Tiger Hill and Mount Tianping in Suzhou, Slender West Lake in Yangzhou, or Orchid Pavilion in Shaoxing. These scenic sites often serve as a community park or the destination of a daytrip.

景园一般择址自然山地或水域区，具开放性和公共性，依托佳山好水建造一些亭阁楼台、寺观庙宇等建筑，形成观景佳地，自身也自成一区而为景园。通常入景园，有序列转折，外接交通，内达景区。

景园规模相对较大，邑郊为多，如苏州虎丘、苏州天平山、扬州瘦西湖、绍兴兰亭等，朝往夕返路程，有城市郊区公园的作用。

Lanting, Shaoxing

Lan (Orchid) Pavilion is located at the foot of Lanzhu Mountain, some 13 km away from the southern suburbs of Shaoxing City. It was originally situated on the southern banks of Lan River in the Song dynasty and relocated to the current site in 1548. The buildings there today were mostly rebuilt in Qing dynasty, which underwent several rounds of restoration.

The park is 200 m deep from north to south and 80 m wide from east to west; its perimeter was once surrounded by paddy fields. Entered from the northern tip, one meanders through a pathway of bamboos before reaching E'chi (Goose Pond). The south of the pond lies an earthen hill forested in greenery, obscuring the main scenery beyond. Moving ahead, one passes through a triangular pavilion to arrive at a rectilinear, single eaves structure; under it stands a monument inscribed with "Lanting" by emperor Kangxi. Turning right, the main scene of the garden is revealed around Liushang (Floating Cup) Pavilion and Youjun Shrine. Liushang Pavilion is built in memory of the floating cup practice, where poets once gathered watching cups filled with yellow wine float past whilst engaged in drinking and poetry writing. Built with four sides, the pavilion differs from others of its kind, as it omitted the sinuous channel altogether. In contrast, the adjacent Youjun Shrine is an exemplary water courtyard. Through the entrance, one arrives at a pool centered by Mohua (Refined Ink) Pavilion. Wrapped by corridors on either side, the walls exhibit a series of calligraphic work including *Lanting xu*; replicated on stone tablets, they recite the floating cup practice from different dynasties. At the end of the pool lie a platform and a three-bays wide hall. Suspended below the hall's ceiling is a portrait of calligrapher WANG Xizhi with a plaque reading "*jinde fengliu*" (attainment of the utmost fluency). The pavements, walls and eaves columns were all sourced with local stone from Shaoxing. In particular, the slender stone columns and slightly arched footbridge before Mohua Pavilion lends the courtyard a graceful elegance.

Lanting is laid out amidst winding paths between lush bamboo groves. The diverted water from Lan River feeds into Goose Pond, meanders through the floating-cup creek before reaching the northern ponds—beyond which flows downstream and finally back into the Yangtze River.

绍兴兰亭

兰亭在绍兴市南郊 13 公里处的兰渚山下。宋代兰亭位于兰溪江南岸山坡，明嘉靖二十七年（1548 年）迁于江北岸现址。现存建筑物均为清代重建，并经后世多次修葺。

兰亭南北进深约 200 余米，东西宽约 80 米，入口在北端，周围为水田。进门经过一段曲折的竹径到达『鹅池』。池南为土山，林木茂密，将兰亭主景部分隐蔽于土山之后。由鹅池屈曲前进，经三角亭达一四角碑亭，单檐盝顶，式样较别致，亭内碑上有康熙御笔所书『兰亭』二字。经此亭折而右，为兰亭主题景区，有流觞亭及右军祠。流觞亭为一纪念亭，其式样作四面厅式，亭内不同于一般流杯亭，不作曲水流觞之举。右军祠是典型水院，进大门两侧为廊，廊内壁上陈列历代《兰亭序》等书法作品的石刻摹本，廊近端是厅堂三间，中悬王羲之画像，匾曰『尽得风流』，堂前为月台。其余院内空间尽为水池，池中方亭『墨华亭』。建筑物铺地、墙体、檐柱等普遍用绍兴当地盛产的石材，建筑所用石柱比例高瘦，具清丽秀美特质，，架于墨华亭上的石板桥微微起拱，造型十分优美。

兰亭布局曲折，竹树森郁，曲水流觞利用兰溪江水引至鹅池，经流杯渠而流至北面诸池再泄于江之下游。

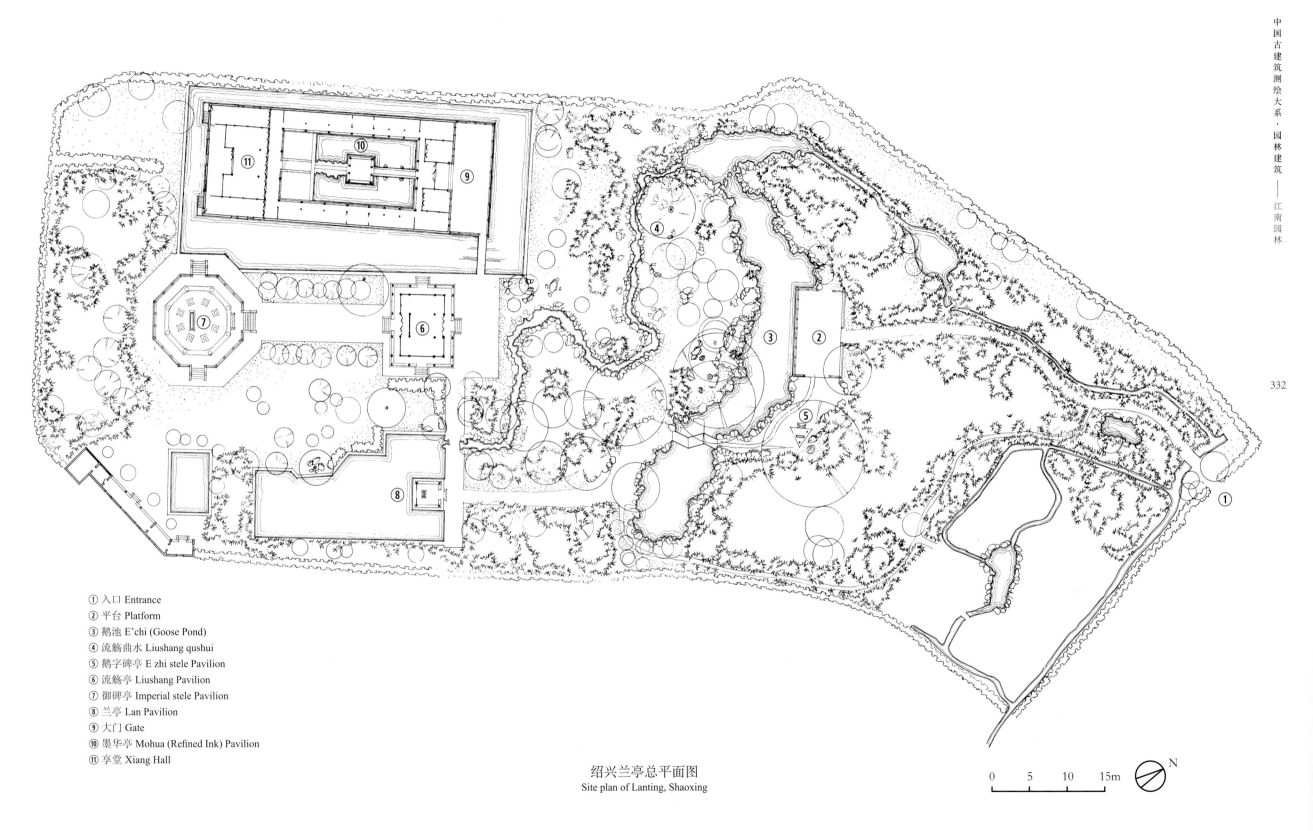

① 入口 Entrance
② 平台 Platform
③ 鹅池 E'chi (Goose Pond)
④ 流觞曲水 Liushang qushui
⑤ 鹅字碑亭 E zhi stele Pavilion
⑥ 流觞亭 Liushang Pavilion
⑦ 御碑亭 Imperial stele Pavilion
⑧ 兰亭 Lan Pavilion
⑨ 大门 Gate
⑩ 墨华亭 Mohua (Refined Ink) Pavilion
⑪ 享堂 Xiang Hall

绍兴兰亭总平面图
Site plan of Lanting, Shaoxing

0 5 10 15m

N

绍兴兰亭透视图
Perspective of Lanting, Shaoxing

流觞亭剖面图
Section of Liushang Pavilion

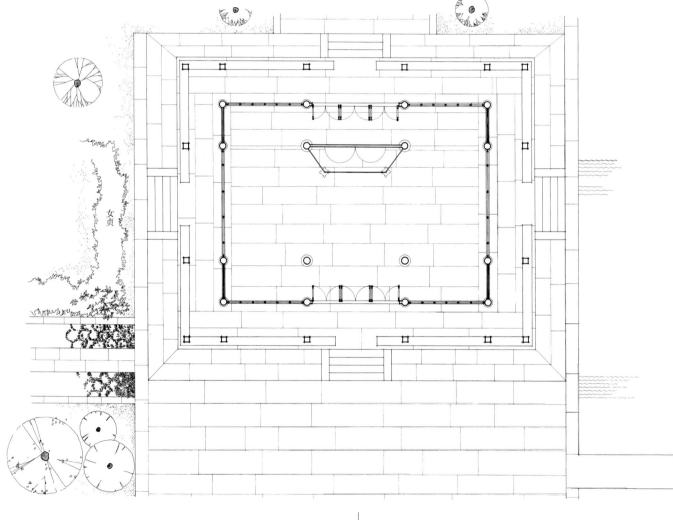

流觞亭平面图
Site plan of Liushang Pavilion

0 1 2m

N

流觞亭立面图
Elevation of Liushang Pavilion

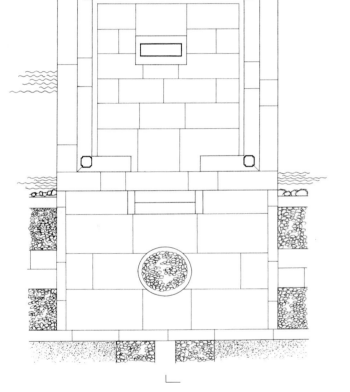

兰亭平面图
Site plan of Lan Pavilion

0 0.5 1 1.5m

N

兰亭剖面图
Section of Lan Pavilion

兰亭立面图
Elevation of Lan Pavilion

Xiaoyingzhou, West Lake, Hangzhou

Xiao Yingzhou is an islet built in 1607 with the excavated silt of West Lake. By 1611, the peripheral embankment had defined the boundary of the inner lakes. Later, zigzag bridges were added, completing the islet in a plan based on the Chinese character for paddy fields (*tian*). The site is known for its unique layout of creating lakes within an islet within West Lake.

The islet is around 300 by 300 m with 65% of its area being water. Willow covers most of the islet, accompanied by other trees such as photinia, maple, dawn redwood, sweet olives, bishopwood, camphor, Chinese wingnut, and yulan trees. From north to south along the bridges, the buildings include Xianxian (Wise Sage) Shrine, a triangular pavilion, a square pavilion, Yingcuixuan (Welcomed Greenery Pavilion), Huaniao (Flower and Bird) Hall, a hexagonal pavilion, Woxin xiangyin (Mutual Affinity) Pavilion and Santan yinyue (Three Pools of Mirrored-moon) Pavilion; dotted by rockeries and bridges, the structures provides several resting spots from the scenic tour.

杭州西湖小瀛洲

小瀛洲岛始建于明万历三十五年（1607年），由浚治西湖时的淤泥堆积而成。万历三十九年（1611年），又周以环形围堤而成岛中有湖的格局。以后又在东西方向连以曲桥，使整个岛的平面呈『田』字形，遂成为别具一格的湖中有岛、岛中有湖的独特布局。

全岛范围约300米×300米，水面约占65%。岛上、堤上遍植柳树，夹以石楠、槭、水杉、桂、重阳木、香樟、枫杨、白玉兰等花木。以南北向桥、岛为轴线，由北向南依次布置先贤祠、三角亭（开网亭）、方亭（亭亭亭）、迎翠轩、花鸟馆、六角亭、我心相印亭、三潭印月等建筑物及湖石峰与曲桥，以供游览休息之用。

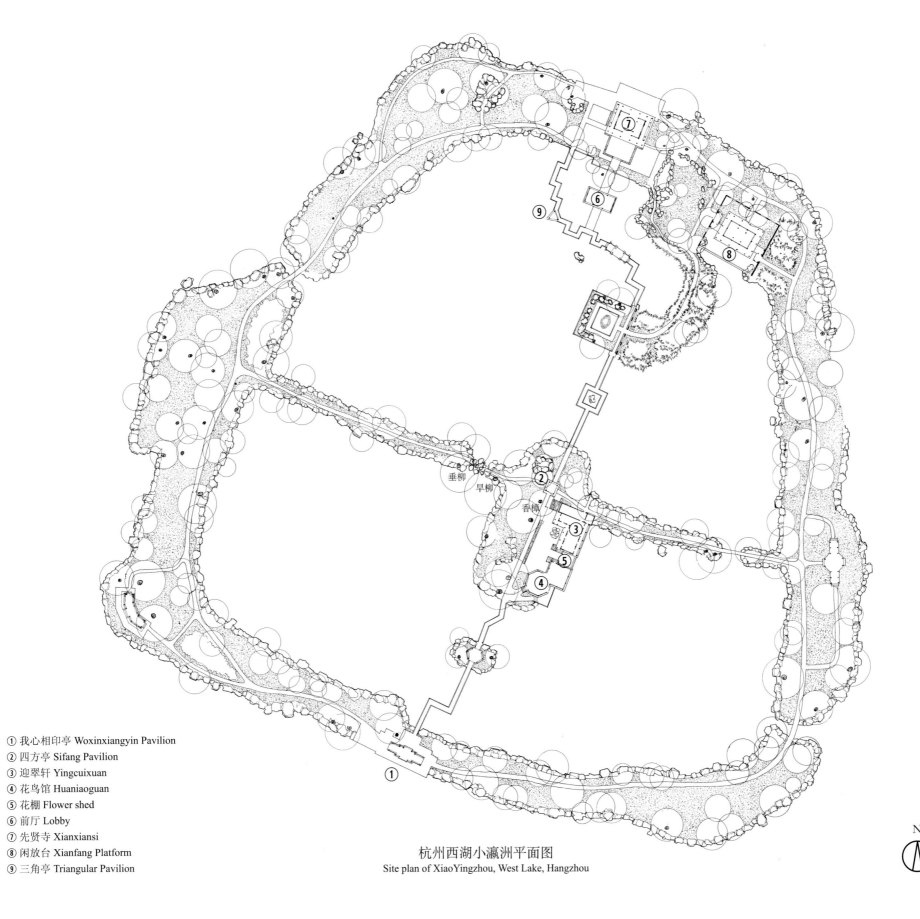

① 我心相印亭 Woxinxiangyin Pavilion
② 四方亭 Sifang Pavilion
③ 迎翠轩 Yingcuixuan
④ 花鸟馆 Huaniaoguan
⑤ 花棚 Flower shed
⑥ 前厅 Lobby
⑦ 先贤寺 Xianxiansi
⑧ 闲放台 Xianfang Platform
⑨ 三角亭 Triangular Pavilion

垂柳
旱柳
香樟

杭州西湖小瀛洲平面图
Site plan of XiaoYingzhou, West Lake, Hangzhou

N

0 10 20 30m

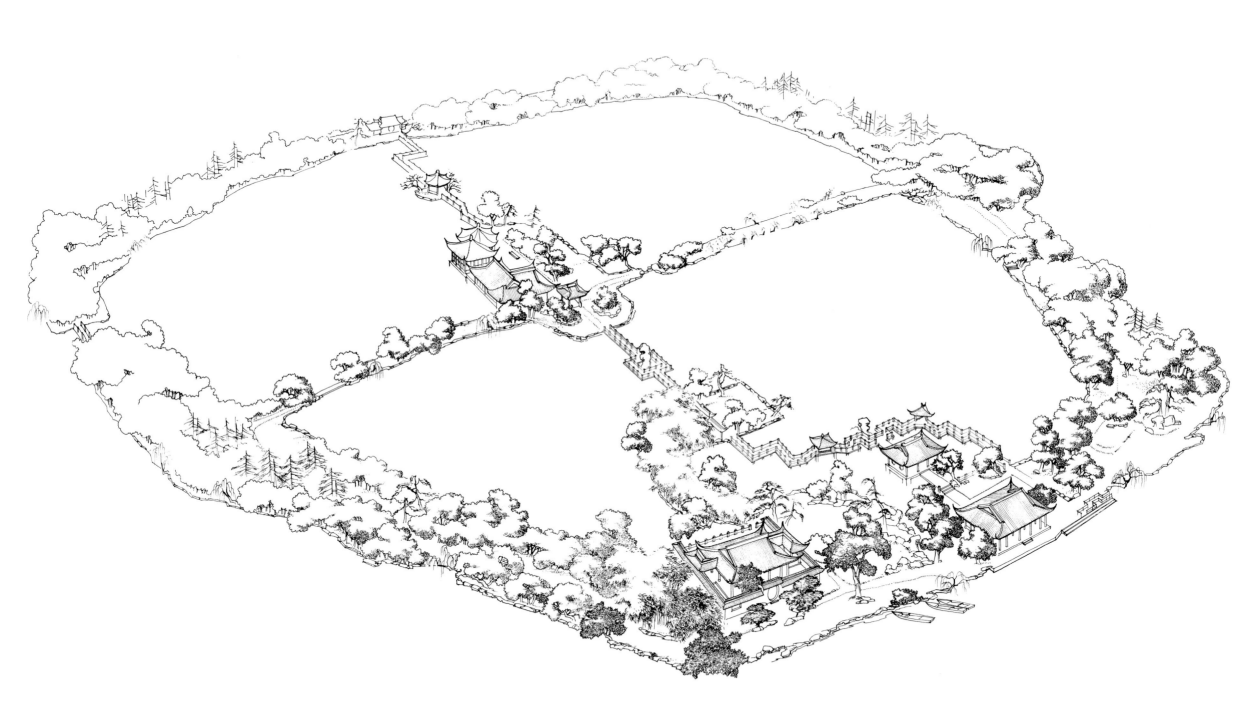

杭州西湖小瀛洲透视图
Perspective of Xiao Yingzhou, West Lake, Hangzhou

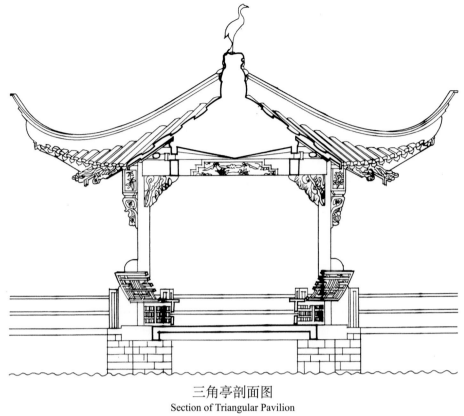

三角亭剖面图
Section of Triangular Pavilion

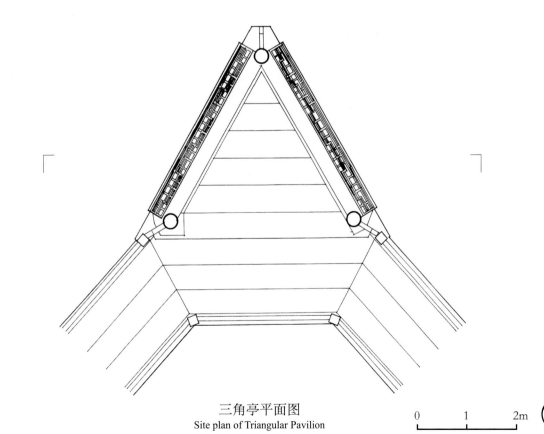

三角亭平面图
Site plan of Triangular Pavilion

0 1 2m

N

三角亭立面图
Elevation of Triangular Pavilion

Xiling Seal Engravers Society, Hangzhou

Founded in 1913, Xiling Seal Engraver Society served as a gathering venue for stone engravers in Zhejiang. Sitting at the foot of Gushan (Gu Hill) in Hangzhou, the property spans to the hilltop, making it a typical mountain garden. Its buildings and sceneries are staged along the irregular slopes with an orderly presence of Baitang (Cypress Hall), Yangxian (Admiration) Pavilion, Shanchan yulou (Mountainous Rain) Study, and Sizhao (Brilliance) Pavilion. The main public venue rests on the hilltop, which exhibits the society's most artistic culminations.

The complex is laid out openly and adapted to its undulating terrain. To the south stands Sizhao Pavilion from which one can behold a view of West Lake. Directly northward across the pond is a cave titled Xiaolong hongdong (Small Tunnel, Deep Cave); bordered by a cliff to its east and a spring between inscribed boulders to its west, the landscape forms a rare hilltop scenery. On the cliff sits Tijin (Expressive) Hall which leads to Huayan Pagoda by walking westward. Its towering figure serves as the focus when viewed from Tijin Hall or Guanle (Observed Joy) Building to its west. At the westmost tip lies Han sanlaoshi (Three Old Rocks of Han) Pavilion.

The garden builds upon the existing landscape in a compact layout filled with diverse sceneries. The spaces are partitioned with the use of buildings, boulders, bamboo and trees—instead of galleries or walls—to make the scenes appear as natural as possible. Meanwhile, inscribed rockeries and statues compliments the landscape. As the spring plunges into the reflective pool beneath, a shimmering scene of water and sky comes to life.

杭州西泠印社

西泠印社成立于1913年，是浙江金石篆刻家的聚会场所，位于杭州孤山，自山趾延亘至山巅，为典型山地园。印社建筑与景物沿山坡布置，依山顺势曲折而上，有柏堂、仰贤亭、山川雨露室、四照阁等，主要活动场所则在山顶，是全社艺术处理最精湛的部分。

山顶平面呈不规则形，空间作开敞式处理。南面有四照阁，凭栏可俯览西湖。阁北隔池直对小龙泓洞，洞东为规印崖，崖下有闲泉，其西又有文泉，凿石聚水，形成难得的山顶泉池。崖上有题襟馆（一名隐闲楼）。自馆而西，跨小龙泓洞，有华严经塔，与观乐楼、题襟馆遥相呼应，为庭园的构图中心。再西为观乐楼，楼南建汉三老石室。

此区庭园在自然山岭基础上进行艺术加工，全局紧凑，丰富多致。在分隔空间方面，交错使用建筑、山岩、竹丛、树木，不用走廊、院墙，使景色自然。同时善用金石与雕像来丰富园景。山顶独得泉水，天光云影映于池中，更为活泼生动。

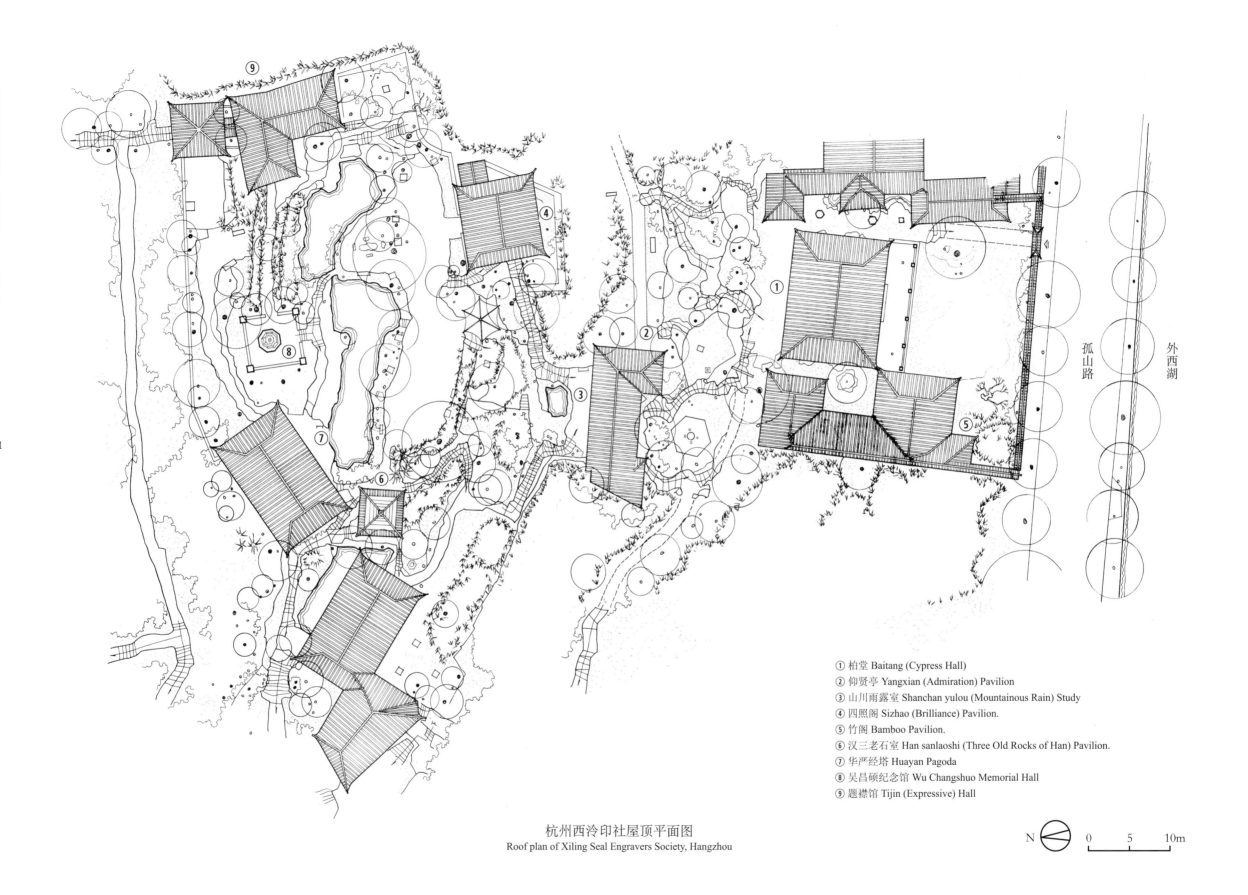

① 柏堂 Baitang (Cypress Hall)
② 仰贤亭 Yangxian (Admiration) Pavilion
③ 山川雨露室 Shanchan yulou (Mountainous Rain) Study
④ 四照阁 Sizhao (Brilliance) Pavilion.
⑤ 竹阁 Bamboo Pavilion.
⑥ 汉三老石室 Han sanlaoshi (Three Old Rocks of Han) Pavilion.
⑦ 华严经塔 Huayan Pagoda
⑧ 吴昌硕纪念馆 Wu Changshuo Memorial Hall
⑨ 题襟馆 Tijin (Expressive) Hall

孤山路

外西湖

杭州西泠印社屋顶平面图
Roof plan of Xiling Seal Engravers Society, Hangzhou

N   0   5   10m

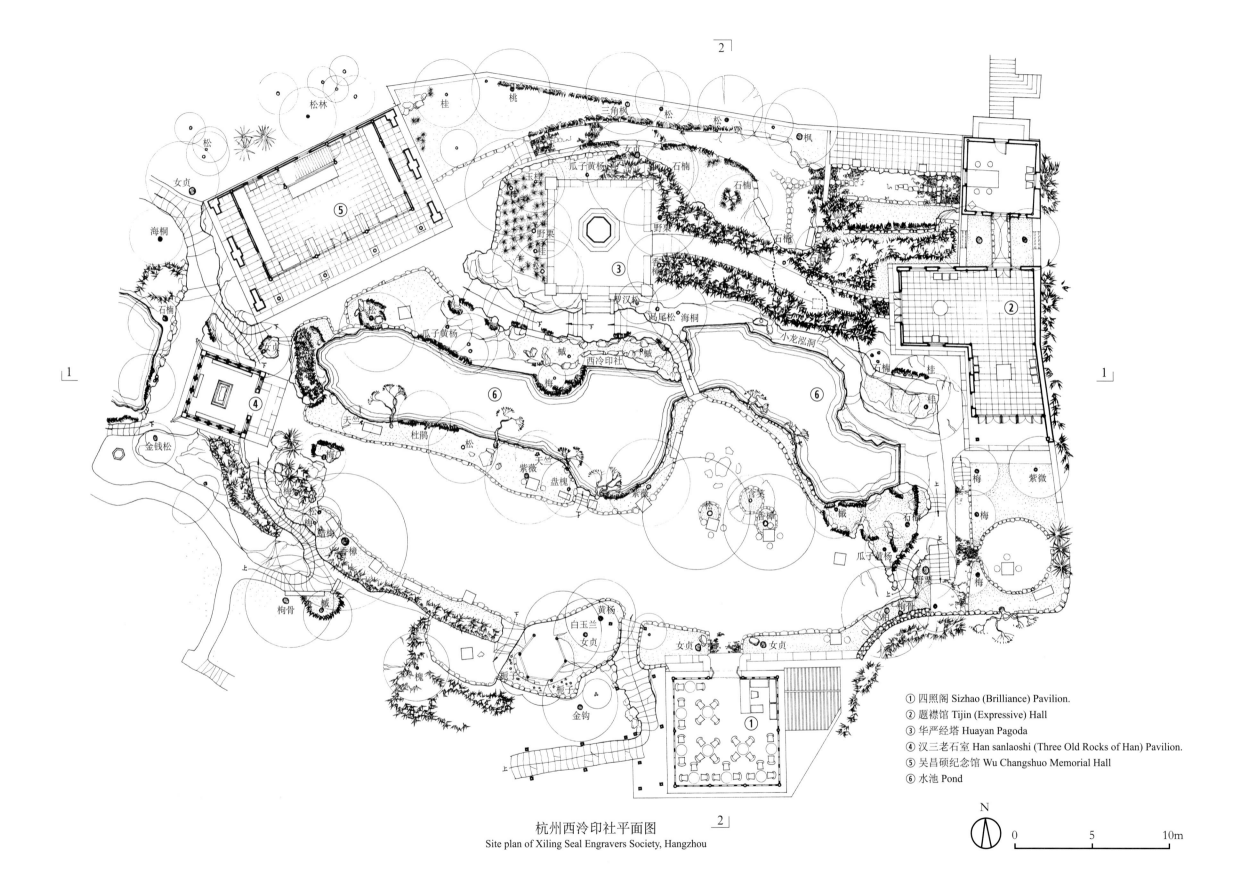

① 四照阁 Sizhao (Brilliance) Pavilion.

② 题襟馆 Tijin (Expressive) Hall

③ 华严经塔 Huayan Pagoda

④ 汉三老石室 Han sanlaoshi (Three Old Rocks of Han) Pavilion.

⑤ 吴昌硕纪念馆 Wu Changshuo Memorial Hall

⑥ 水池 Pond

杭州西泠印社平面图
Site plan of Xiling Seal Engravers Society, Hangzhou

N

0   5   10m

杭州西泠印社 1—1 剖面图
Section 1-1 of Xiling Seal Engravers Society, Hangzhou

0 1 2 3m

杭州西泠印社 2-2 剖面图
Section 2-2 of Xiling Seal Engravers Society, Hangzhou

0 1 2 3m

Yanyulou, Jiaxing

Yanyulou (Misty Rain Building) is the main structure on an island in South Lake in Jiaxing, Zhejiang. Covering an area of 17 *mu*, the building's name also refers to the island's garden as a whole.

In 1548, a magistrate named ZHAO Ying ordered the dumping of silt dredged from the city's moat construction into South Lake to form an island. Later, Yanyu Building was erected on the island; its name references the original lakeside building used by QIAN Yuanliao, a son of the emperor of Wu-yue kingdom from the Five Dynasties period. After Qing dynasty, new buildings including Qinghui (Sunshine) Hall, Guyunyi (Lone Cloud Pavilion), and Xiaopenglai were added to highlight the presence of Yanyu Building. Comprised of pavilions, buildings, rockeries, galleries, and ancient inscriptions, the complex emulates the qualities of a Jiangnan Garden. Whether viewed from far or up close, Yanyu Building stands out from surrounding structures without appearing isolated, reflecting the success of its layout.

Yanyu Building was abandoned several times over the course of its life. The one people see today was rebuilt during the the Republic of China. At two-stories tall, the building is oriented south; aside from the platform in front, no structure obstructs its view from South Lake. The most mesmerizing scene is when distant boats and nearby trees fades into the mist of drizzling rain, as if the view was shrouded in silk—hence its name Yanyu. It provides a focus amidst the lake.

Behind Yanyu Building lies an open courtyard of an irregular layout. Framed on three sides by buildings and galleries, the courtyard is filled with rockeries, trees and flowers. In addition, the periphery is lined with walls to its south, west and north; the resulting space is one of a serene, secluded ambiance without the feeling of desolation. Two ancient ginkgo trees marks the front of the building, accompanied by other old trees around it and courtyard; these include hackberry, Chinese cinnamon, white yulan, *mudan* (tree peony), cherry, wintersweet, camphor, maple, Chinese scholar tree, elm, and loquat. Amidst all the greenery, Yanyu Building appears to be tucked between the foliage of a deep forest.

嘉兴烟雨楼

嘉兴南湖湖心岛位于南湖中心，面积约 17 亩，烟雨楼为岛上主要建筑，亦以此泛指岛上园林。

明嘉靖二十七年（1548 年），知府赵瀛疏浚城河，将余土运入湖中填成小岛，并于岛上建烟雨楼，其名沿用自五代吴越国中吴节度使广陵王钱元璙在湖滨所筑之楼。清后又相继建成清晖堂、孤云簃、小蓬莱、来许亭、鉴亭、宝梅亭、东西御碑亭、访踪亭等建筑，烘托烟雨楼，形成了以楼为主体的园林建筑群。亭台楼阁、假山回廊、古树碑刻，错落有致，是典型的江南园林。无论从湖滨远望还是从长堤近观，此楼都是前拥后护，左辅右弼，主导地位突出而又不陷于孤立无依，建筑布局十分成功。

烟雨楼几经兴废，现存为民国年间重建。楼作二层，面南。前面除平台外，别无其他建筑遮挡，视野开阔，南湖景色，尽收眼底。最动人景色是细雨迷蒙之时，湖上远舟近树如笼于轻纱之中，故以『烟雨』命名此楼，起到画龙点睛作用。

楼后是一不规则开阔庭院，三面环以屋宇及游廊，院中堆假山，植花木，内容较丰富。楼的南、西、北面，加筑墙一周，故楼下空间更加封闭，使处于湖中岛屿免除了空旷无依的感觉，增加了安全、宁静的意趣。楼前两株古银杏树，楼周有多株古朴树，如桂、白玉兰、牡丹、樱桃、蜡梅、香樟、槭、盘槐、榆、枇杷等，分布楼周及庭院中，使烟雨楼处在郁郁葱葱的林木环境之中。

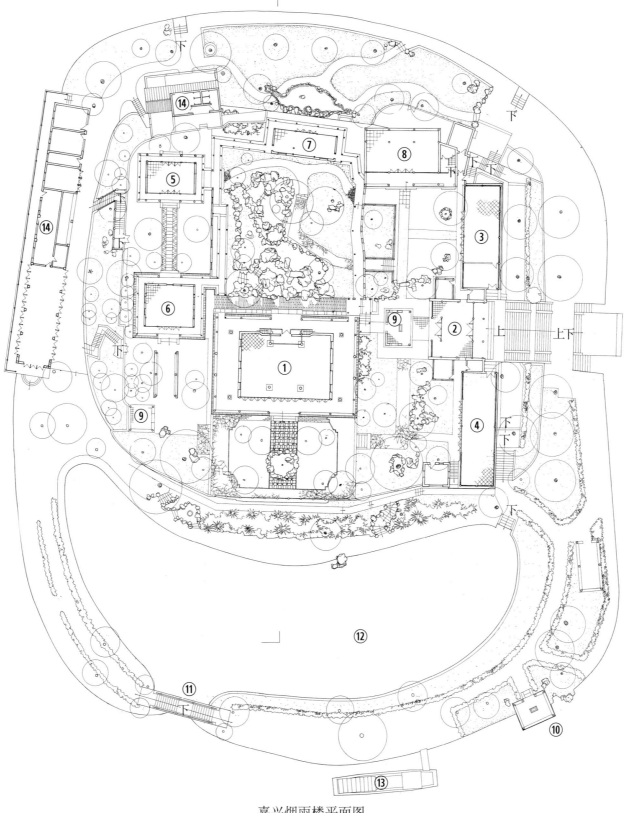

① 烟雨楼 Yanyulou (Misty Rain Building)
② 清晖堂 Qinghui (Sunshine) Hall
③ 孤云簃 Guyunyi (Lone Cloud Pavilion)
④ 小蓬莱 Xiaopenglai
⑤ 来许亭 Laixu Pavilion
⑥ 鉴亭 Jian Pavilion
⑦ 宝梅亭 Baomei Pavilion
⑧ 观音阁 Guanyin Pavilion
⑨ 碑亭 Stele Pavilion
⑩ 访踪亭 Fangzong Pavilion
⑪ 石桥 Stone bridge
⑫ 水面 The Water
⑬ 纪念船 Memorial ship
⑭ 服务用房 Service room

嘉兴烟雨楼平面图
Site plan of Yanyulou, Jiaxing

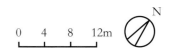

0   4   8   12m

N

朴树

蜡梅

朴树

朴树

⑦

棕榈

梧桐

朴树

金桂

樱桃

⑤

小叶黄杨

香樟

朴树

马尾松

银桂

棕榈

蜡梅

罗汉松

罗汉松

红枫

牡丹

⑥

罗汉松

朴树

小叶黄杨

盘槐

楠木

蜡梅

②

枇杷

榆树

①

枇杷

槭树

柏

松

棕榈

枇杷

蜡梅

枇杷

④

⑧

女贞

白玉兰

桂花

桂花

白玉兰

银杏

银杏

橘树

钓鳌矶

香樟

棕榈

① 烟雨楼 Yanyulou (Misty Rain Building)
② 清晖堂 Qinghui (Sunshine) Hall
③ 孤云簃 Guyunyi (Lone Cloud Pavilion)
④ 小蓬莱 Xiaopenglai
⑤ 来许亭 Laixu Pavilion
⑥ 鉴亭 Jian Pavilion
⑦ 宝梅亭 Baomei Pavilion
⑧ 碑亭 Stele Pavilion

嘉兴烟雨楼庭院平面图
Courtyard plan of Yanyulou, Jiaxing

0 1 2m

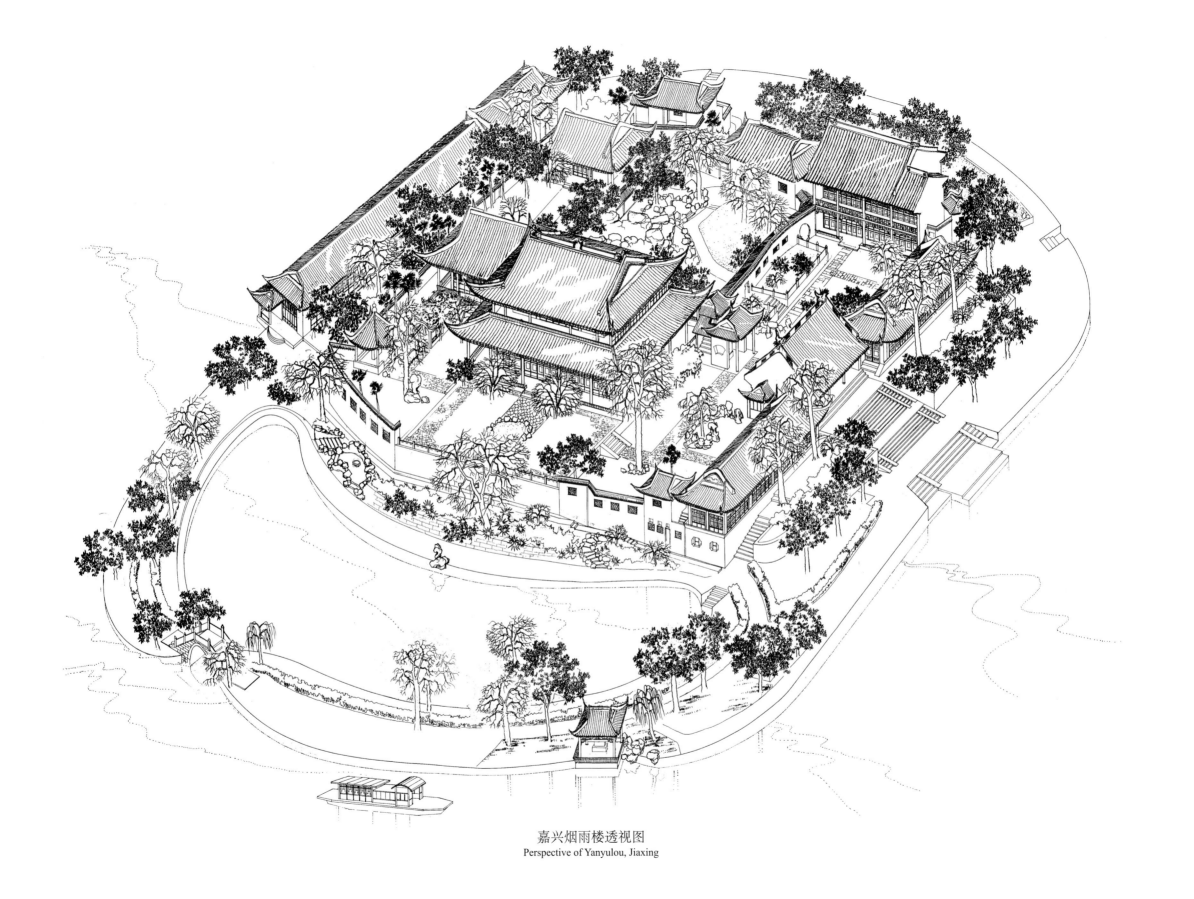

嘉兴烟雨楼透视图

Perspective of Yanyulou, Jiaxing

嘉兴烟雨楼东北立面图
Northeasten elevation of Yanyulou, Jiaxing

0　2　4　6m

嘉兴烟雨楼剖面图
Section of Yanyulou, Jiaxing

0　2　4　6m

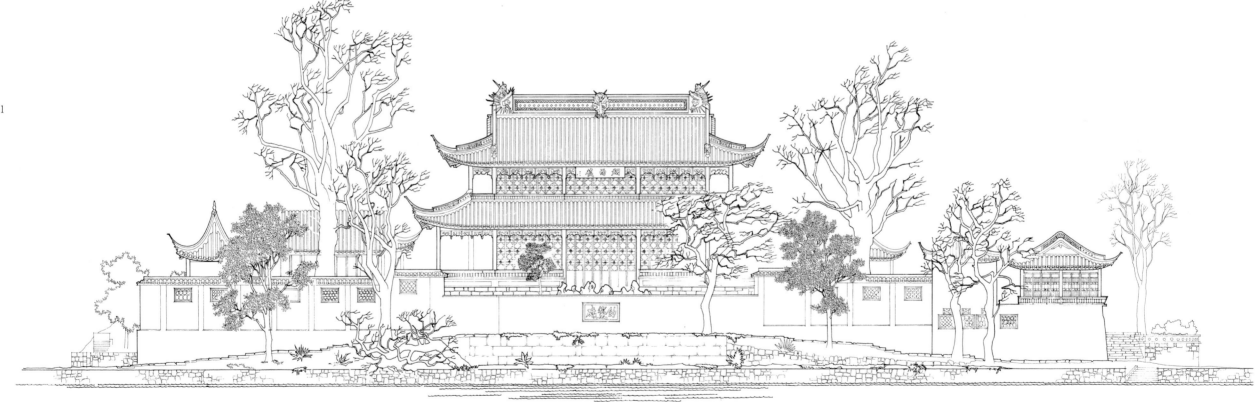

嘉兴烟雨楼东南立面图
Southeasten elevation of Yanyulou, Jiaxing

0　1　2　3m

Tiger Hill, Suzhou

Huqiu, or Tiger Hill, is located 3.5 kilom away in the northwestern part of Suzhou; it is also known as Haiyongqiu (Sea Surge Hill). The site initially held cottages built by the grandsons of WANG Dao, a minister from Eastern Jin dynasty. Later, two monasteries were added behind the residence's east and west; this marked the beginning of Tiger Hill Monastery. Today, aside from Yunyan (Cloud Boulder) Pagoda from early Song dynasty and Ershanmen (Second Hill Gate) from Yuan dynasty, the rest of the buildings were built after the Taiping Rebellion. In fact, most buildings were never restored, such as the gate of the central hall, the main hall, and the rear hall. The northwest of Tiger Hill lies the highest peak; a secondary peak stretches along its southeast. In between lies an expansive rock field called Qianrenzuo (Thousand People Seat) and the steep cliffs of Jianchi (Sword Pond). While the west has more buildings on raised terrains, the east is relatively flat with fewer buildings.

Tiger Hill Monastery's layout is reflective of its undulating terrain. Starting from Toushanmen (First Hill Gate) on Shantang Street, one ascends along an upward axis through Ershanmen, Yongcui (Embraced Greenery) Residence, Gonsheng jiangtai (Lecture Platform) and Qianrenzuo. Shifting eastward, a flight of 53 stone steps leads up to the monastic front gate, through which one arrives at the forecourt of the main hall. To the east stands a double eaves hall, behind which rests the seven-storied pagoda, followed by a rear hall. In Ming dynasty, a pavilion was initially planned on the left flank of the main hall. However, the irregular terrain forced the building to rotate 90 degrees from the first and second gates' axis with longer sides facing east and west. Not only did this adapt to the landform but it effectively resolved the layout of the courtyard and the halls.

The western terrain is taller and steeper, on which stands a range of buildings including Lengxiang (Cold Fragrance) Pavilion, Zhishuang (Cooling) Pavilion, Stone Guanyin Hall, Lu Yu Stone Well, etc. In contrast, the flatter eastern terrains have fewer buildings remaining.

苏州虎丘

虎丘在苏州西北 3.5 公里处，又名『海涌山』。东晋王导之孙司徒王珣、司空王珉在山上建别墅，后舍宅为东西二寺，是虎丘佛寺之始。现寺内建筑除云岩寺塔为宋初所建，二山门为元代遗构外，其余皆太平天国以后所建。其中殿门、大殿及后殿等建筑始终没有恢复。虎丘西北为主峰，有二岗向东南伸展，二岗之间有平坦石场（『千人坐』）及剑池岩壑。西岗地势较高，现存建筑较多。东岗地势较平，现存建筑无多。

虎丘寺庙布局依山就势，从山塘街头山门起，沿轴线而进，过二山门，有拥翠山庄、生公讲台、千人坐；由千人坐东登 53 级石级，方入寺院正门。正门朝南，入门为大殿前院，东侧有重檐殿宇，大殿后有七级宝塔，再有后殿，明代又在正殿前左侧正对门建有一阁。由于山顶地形缘故，殿宇轴线转而为东西向，与头山门、二山门之轴线成 90° 角相交，既结合地形，又解决了殿庭的布局。

西面山岗地势较高，建筑有冷香阁、致爽阁、石观音殿、陆羽石井等；东面地势较平，所存建筑不多。

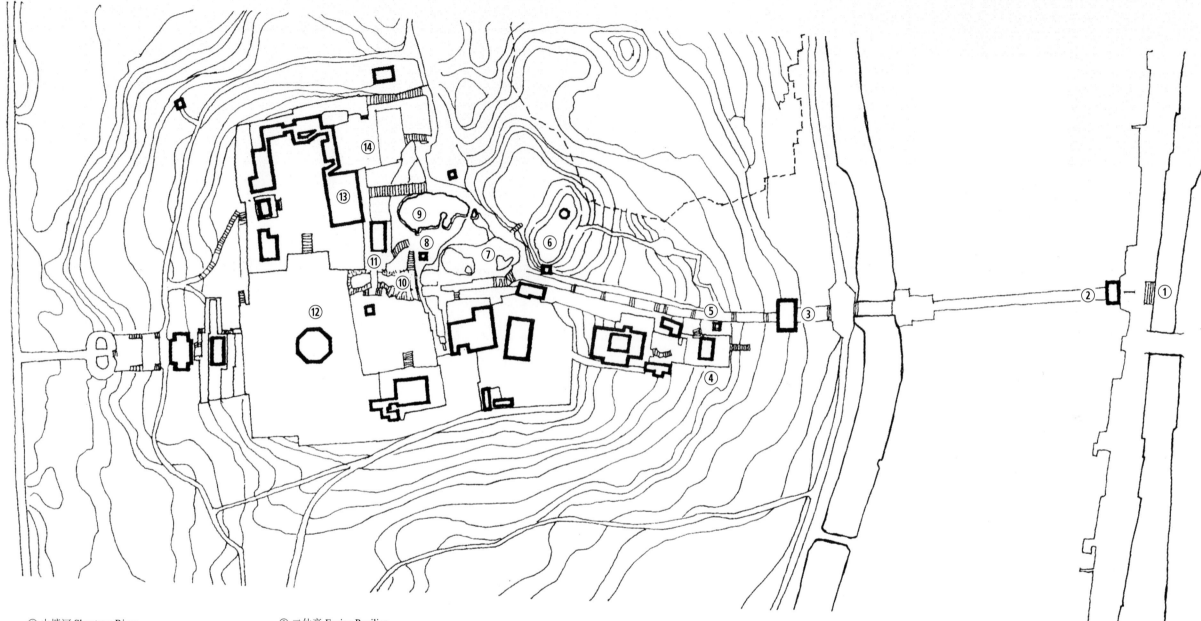

① 山塘河 Shantang River
② 头山门 Toushanmen (First Hill Gate)
③ 二山门 Ershanmen (Second Hill Gate)
④ 拥翠山庄 Yongcui (Embraced Greenery) Residence
⑤ 憨憨泉 Hanhanquan
⑥ 真娘墓 Zhenniang Tomb
⑦ 千人坐 Qianrenzuo (Thousand People Seat)
⑧ 二仙亭 Erxian Pavilion
⑨ 白莲池 Bailian Pond
⑩ 剑池 Jianchi (Sword Pond)
⑪ 悟石轩 Wushixuan
⑫ 云岩寺塔 Yunyan (Cloud Boulder) Pagoda
⑬ 天王殿 Tainwang Hall
⑭ 屐石 Jishi

苏州虎丘总平面图
Site plan of Tiger Hill, Suzhou

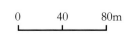

0 40 80m

N

Wushi Pavilion rests on the hillside of Tiger Hill. It was conceived as an enlarged gallery leading to the hall to its right, used mainly as a resting spot. Beneath the pavilion's south is a stretch of whitewashed wall, setting a dark-bright contrast against the natural rockeries and long windows above. The interior is plainly furnished, opening out to a distant view of the sky and foliage beyond.

虎丘悟石轩位于虎丘半山腰处。轩由通向右边殿堂的通廊扩大空间而成，作憩息之处。南面下部用大片粉墙，与自然山石及上部长条窗形成强烈明暗对比，具有典雅的天然之趣。室内家具陈设纯朴。内外空间开阔，极目远眺，碧空绿荫尽收眼底。

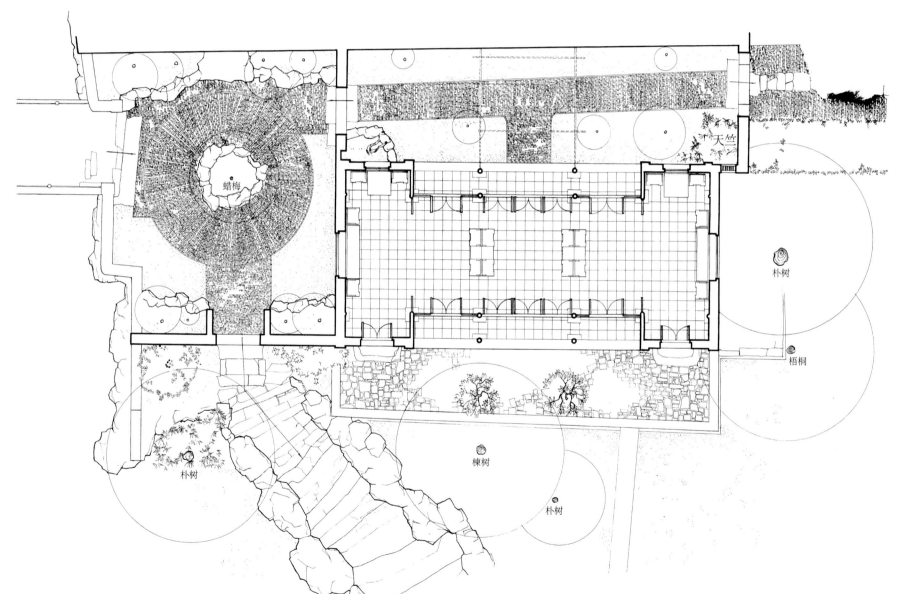

蜡梅

天竺

朴树

梧桐

楝树

朴树

朴树

N

0　1　2　3m

悟石轩平面图
Site plan of Wushixuan

悟石轩剖透视图
Sectional perspective of Wushixuan

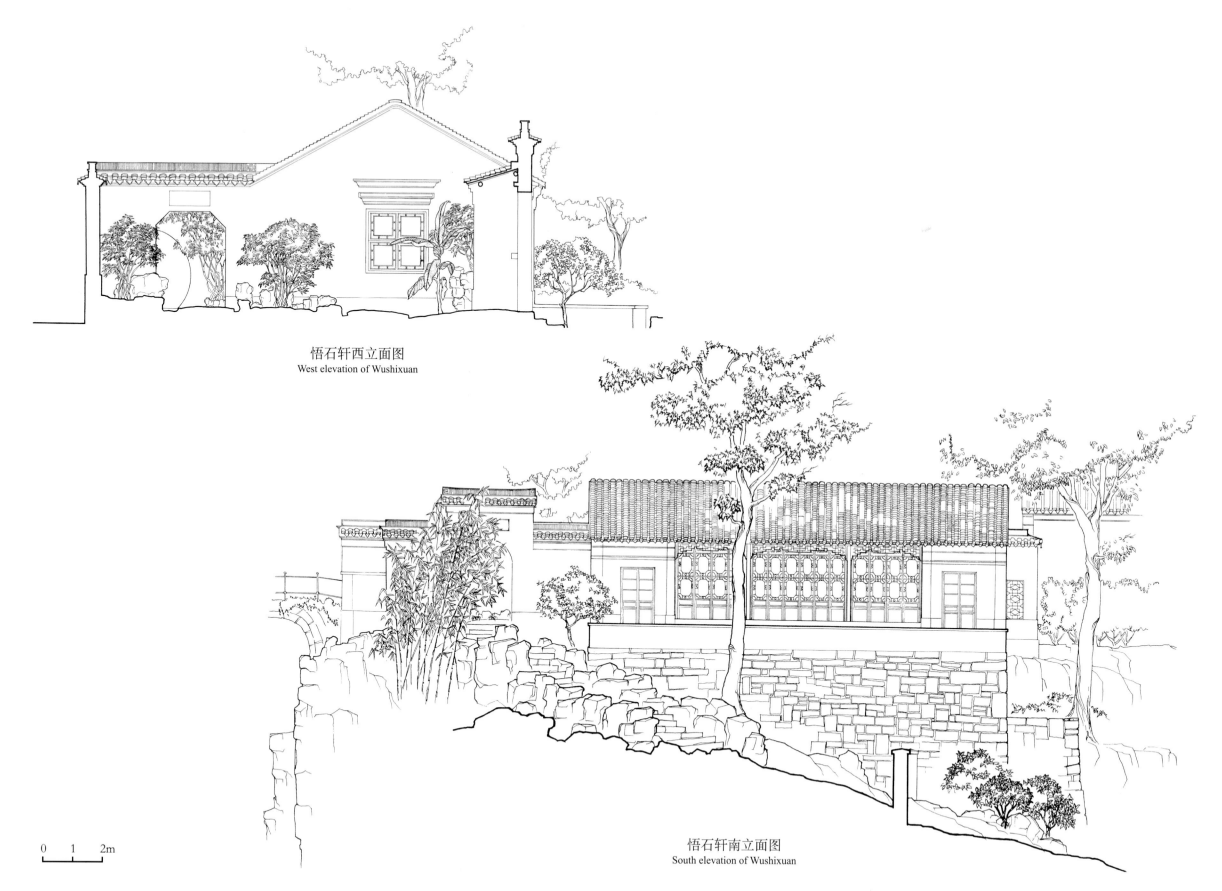

悟石轩西立面图
West elevation of Wushixuan

悟石轩南立面图
South elevation of Wushixuan

0 1 2m

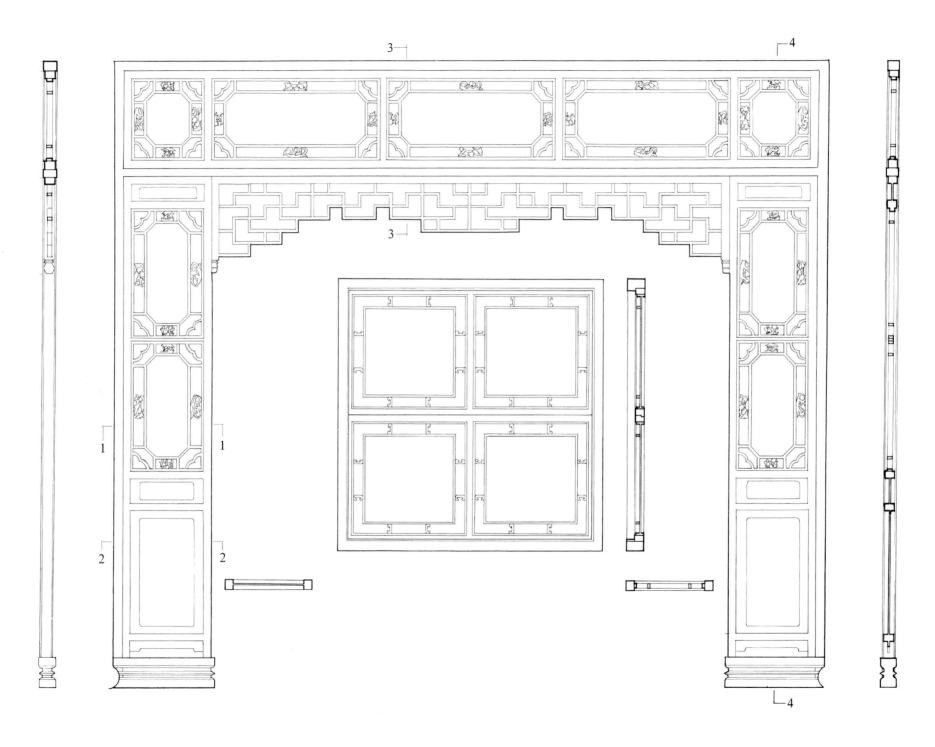

装修大样图
Interior partitions

0   0.5m

Slender West Lake, Yangzhou

Slender West Lake is located in the northwest section of Yangzhou's old city from the Ming and Qing dynasties. At 5,000 m long, the lake runs from an ancient bridge from the city's southern gate to Pingshan Hall in the north. It partially served as a moat and other river courses built in Tang and Song dynasties. In Qing dynasty, Yangzhou enjoyed a prosperous economy from its well-developed salt industry. In addition, the city also served as a stop for the southern expeditions taken by emperors Qianlong and Kangxi, which furthered its development in landscape gardens. From the imperial dock outside of Tianning Monastery to Pingshan Hall in the north, all the buildings line up to form a continuous scenery along the canal; this formation constitutes the basic layout of Slender West Lake. The site was officially added to the World Heritage List in 2014.

扬州瘦西湖

瘦西湖在扬州明清老城西北部，南自南门古渡桥，北至蜀岗平山堂，全长 5000 多米，是在唐宋时的部分护城河和其他河道基础上经历代浚治而成。清代扬州盐业发达，经济繁盛，加上康乾下江南的驻足需求，出现了瘦西湖畔园林建设高潮，城北自天宁寺行宫门前御码头开始抵平山堂，沿河两岸楼台首尾相接，形成曲对水上游线连续展开的园林群，为瘦西湖之基本格局。2014 年被列入世界文化遗产名录。

① 天宁寺 Tianning Monastery
② 冶春园 Yechun Garden
③ 北门桥 Beimen Bridge
④ 卷石洞天 Juanshi dongtian
⑤ 红桥 Red Bridge
⑥ 四桥烟雨 Siqiao yanyu
⑦ 徐园 Xu Garden
⑧ 小金山 Xiaojinshan (Small Gold Mountain)
⑨ 凫庄 Fuzhuang
⑩ 五亭桥 Witing Bridge (Five Pavilion Bridge)
⑪ 白塔 White Pagoda
⑫ 莲性寺 Lianxing Monastery
⑬ 二十四桥 The twenty fourth Bridge
⑭ 熙春台 Xichuntai
⑮ 平山堂 Pingshan Hall
⑯ 西园 West Garden
⑰ 大明寺 Daming Monastery
⑱ 观音山 Guanyin Mountain
⑲ 唐城遗址 Tang city site
⑳ 扬州城区 Yangzhou City

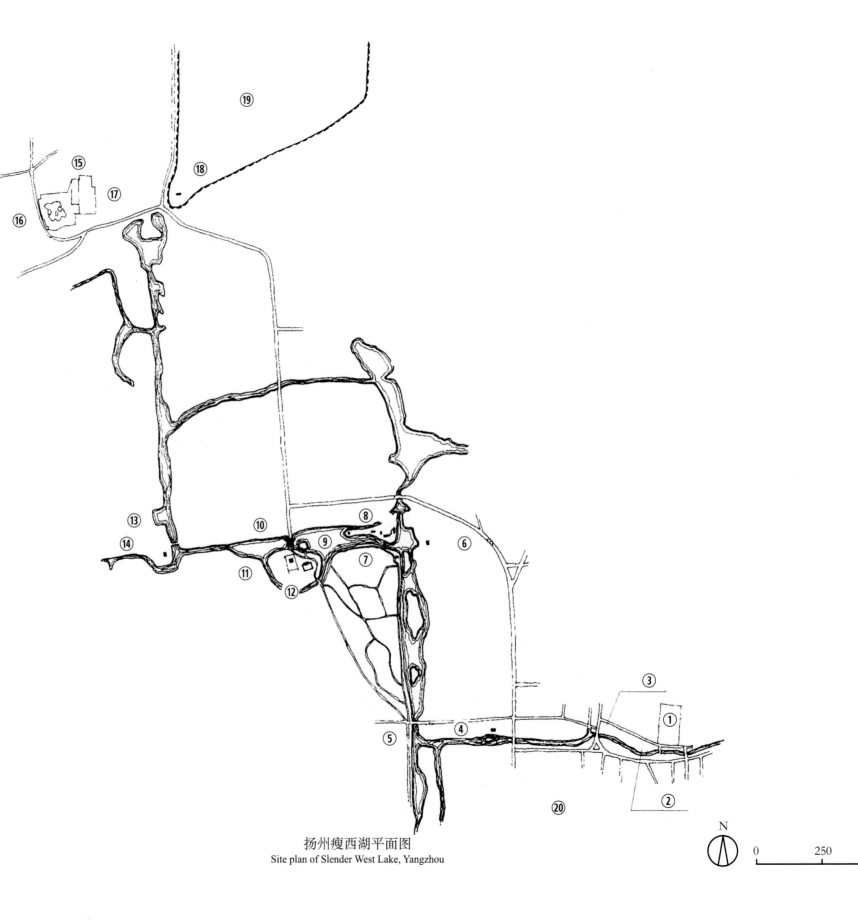

扬州瘦西湖平面图
Site plan of Slender West Lake, Yangzhou

N

0   250   500m

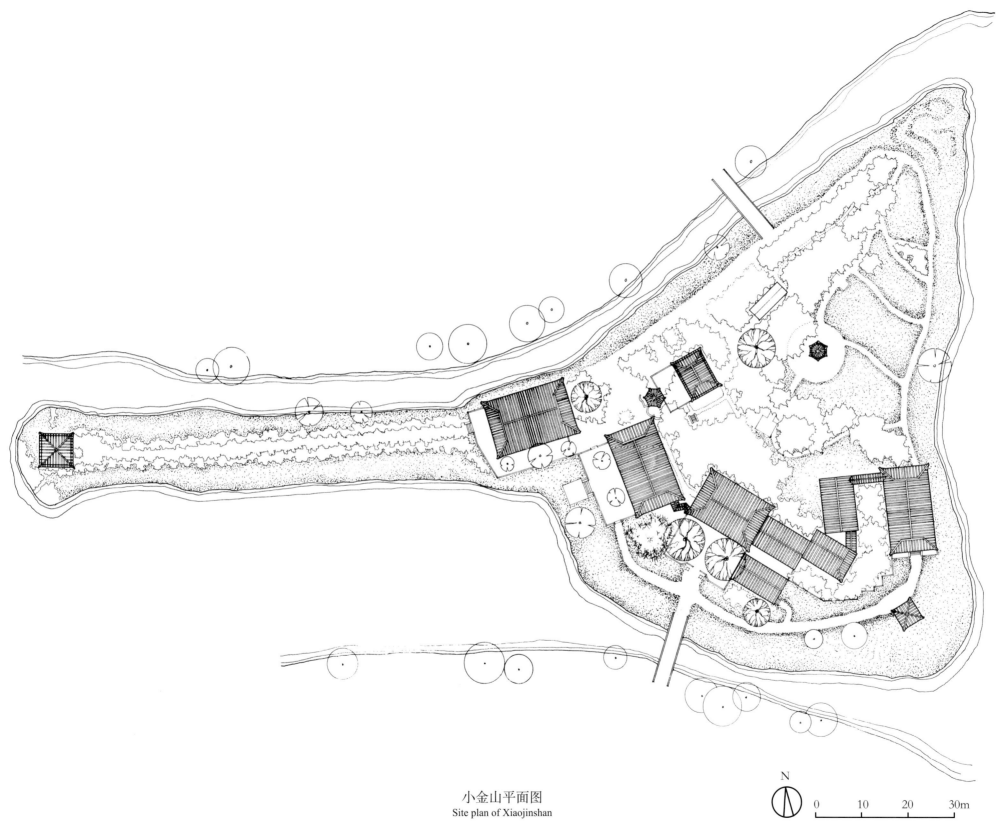

小金山平面图
Site plan of Xiaojinshan

N

0　　10　　20　　30m

五亭桥
Wuting Bridge

Wuting Bridge, or Five Pavilion Bridge, was built in 1757 across the lake from Xiaojinshan (Small Gold Mountain). Its form is comprised of five pavilions sitting on a stone bridge, punctuated by arches of various size. Its stylistic diversity presents a unique focus for the surrounding scenery.

五亭桥与小金山隔湖相望，建于乾隆二十二年（1757年）。桥上有五亭，亭下由大小不同的拱券组成桥身，构思奇特，造型极为丰富。

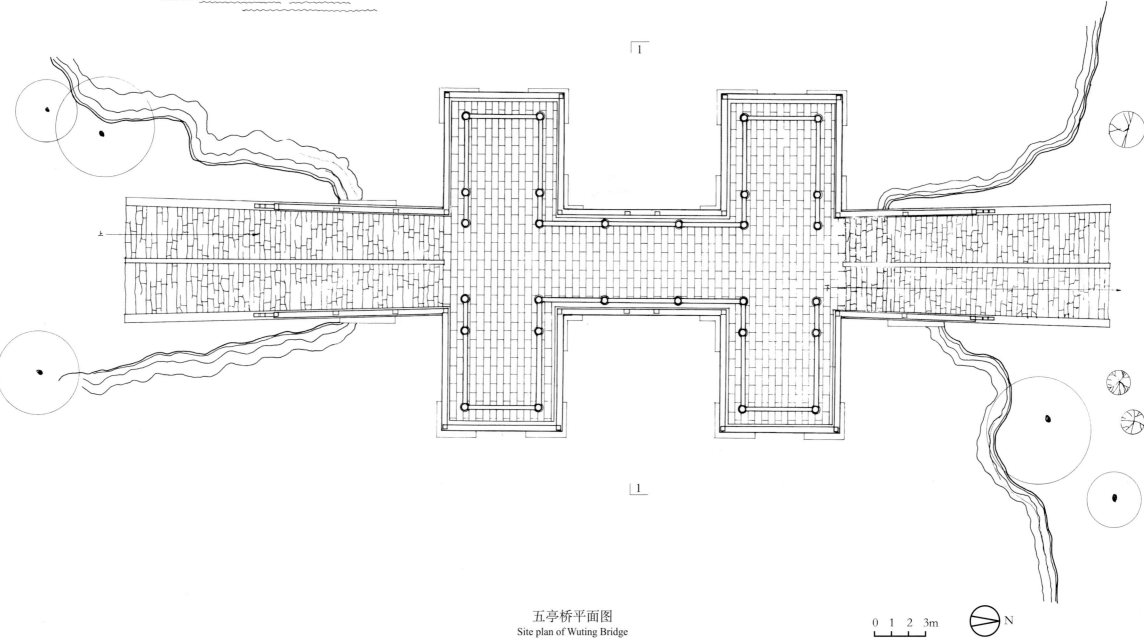

五亭桥平面图
Site plan of Wuting Bridge

0 1 2 3m

N

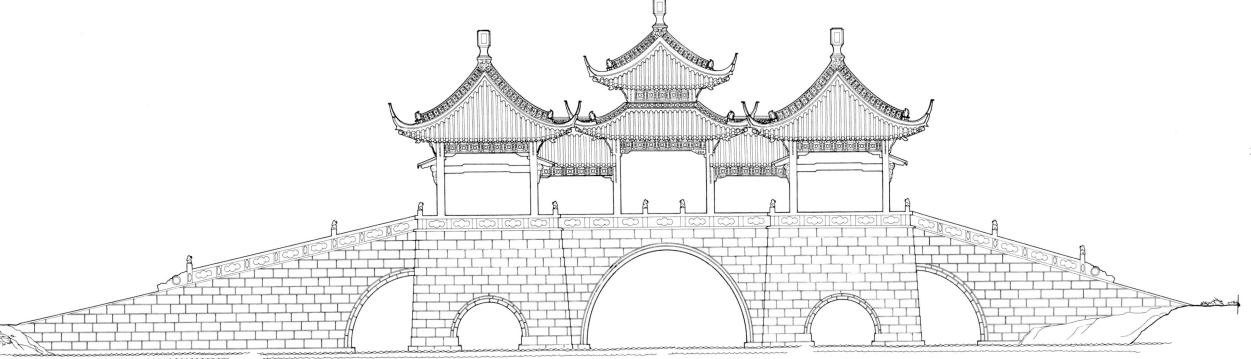

五亭桥东立面图
East elevation of Wuting Bridge

0　1　2　3m

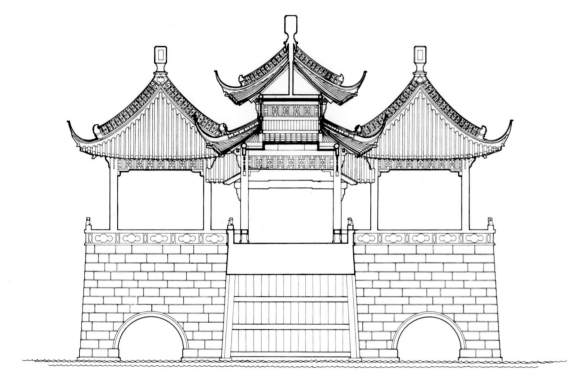

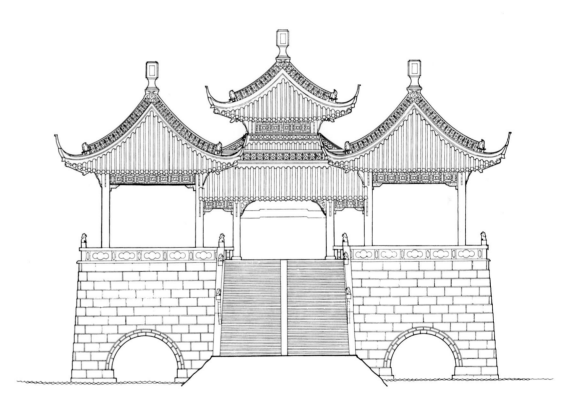

五亭桥 1—1 剖面图
Section 1-1 of Wuting Bridge

五亭桥南立面图
North elevation of Wuting Bridge

0　1　2　3m

Baita, or White Pagoda, is a stupa within the premise of Lianxing Monastery, south of Five Pavilion Bridge. It is based on the building of the same name from Beijing's Beihai Park, though much smaller in scale. Its color, profile, and height prevail amongst the lakeside buildings of Slender West Lake—serving as the final touch to completing the scenic site.

白塔位于五亭桥南的莲性寺中，系仿北京北海白塔所建，但规模小得多。其色彩、形象、高度在瘦西湖各建筑中最为突出，故能起到画龙点睛和统领全局的作用。

白塔
White Pagoda

白塔南立面图
North elevation of White Pagoda

白塔平面图
Site plan of White Pagoda

N

0   2   4m

说明

一、关于图版：

本书图版辑录东南大学（原南京工学院）建筑学院自20世纪50年代起至今的传统园林景观测绘图纸，经整理加工而成。图版包括：《苏州古典园林》（刘敦桢著，中国建筑工业出版社，1979年10月第一版）、《江南理景艺术》（潘谷西编著，东南大学出版社2001年4月第一版）、《江南园林图录——庭院·景观建筑》（刘先觉、潘谷西合编，东南大学出版社2007年9月第一版）的部分图纸，以及2007年硕士研究生测绘、2015—2017年四年级建筑学与风景园林本科生测绘的成果。

二、关于编写：

体例结构由陈薇设计和分类；具体园林内容介绍由是霏、史文娟整理完成，陈薇校核和修改；图版处理：是霏、史文娟、郑婳、高雯汐、谢菲；审核：陈薇；英文翻译：荷雅丽、周彦邦、邹春申。

三、关于工作：

（一）《苏州古典园林》：

1953年，刘敦桢先生组织南京工学院与华东工业建筑设计院合办的中国建筑研究室的人员，对苏州古典园林作了普查。后又领导南京工学院建筑系原建筑历史教研组与原建筑工程部建筑科学研究院建筑理论与历史研究室南京分室工作人员对苏州重点园林进一步作了调查研究和测绘，1960年撰写《苏州古典园林》初稿，1963年进行修改和补充。

参加该书稿研究、测绘、制图、摄影的主要成员有：朱鸣泉、沈国尧、詹永伟、金启英、傅高杰与原建筑工程部建筑科学研究院建筑理论与历史研究室南京分室的有关人员和南京工学院建筑系的教师、同学。

1973年后整理书稿的人员有：潘谷西、刘先觉、乐卫忠、郭湖生、叶菊华、刘叙杰。

（二）《江南园林图录》：

1963年6—7月，南京工学院建筑系1966届本科生测绘苏州、无锡、扬州庭院与园林建筑实例。指导教师：潘谷西、刘先觉、杜顺宝、陈湘、李伯先、徐定樑。

1963年12月，南京工学院建筑系1966届本科部分学生测绘了南通、杭州庭院与园林建筑实例。指导教师：潘谷西、刘先觉、杜顺宝、陈湘等。

1964年7月，南京工学院建筑系1966届本科部分学生及教师绘制修改测绘图。指导教师：潘谷西。

（三）《江南理景艺术》的相关补充：

1981年5月，潘谷西、杜顺宝、何建中测绘无锡寄畅园平面、剖面图。

1982年5月，南京工学院建筑系1982届本科毕业设计小组测绘洞庭西山庭院实例。指导教师：潘谷西。

1982年6月，南京工学院建筑系1983届本科生测绘扬州、嘉兴、杭州园林实例。指导教师：杜顺宝、庄金元、郑光复等。

1996年，东南大学李浈、查群、章忠民测绘绮园、烟雨楼。指导教师：朱光亚。其中绮园平面图原稿由浙江省文物考古研究所提供。

1996年8月，东南大学建筑系1997届部分学生测绘南京、泰州园林实例。指导教师：丁宏伟、陈薇、李海清。

1997年6月，东南大学郭华瑜、查群测绘退思园。

1997年9月，东南大学1993届学生补测南京瞻园平面。指导教师：成玉宁。

1997年9月，胡石、冯炜补测杭州郭庄平面图。

（四）近年测绘补充：

2007年11月，测绘泰州乔园。硕士研究生：杨俊、薛垲、祁昭。指导教师：陈薇。

2015年8—9月，测绘苏州艺圃。2012级本科生：张悦、钱志达、邵建东、徐文婷。硕士研究生：刘硕、黄宇星、左亚男、解文慧、夏意、盛苏晨、宋佳、丰远、朱孟楠、严恒。

2016年8—9月，测绘苏州耦园。2013级本科生：华正晨、于广洲、杨一鸣、张亚、吴江源、马力凌、马琳、陈爽、戴文翼、陶博双、施俊婕、高雪、崔吉吕、李灏、刘晓雯。硕士研究生：叶枝、郭逸文、孟阳、高文娟。指导教师：陈薇、贾亭立、是霏。

2017年8—9月，测绘苏州沧浪亭。2014级本科生：杨帆、徐闻、国佳琪、高雯汐、怀玉菡、商宁悦、郑婳、陆倚竹、陈豪、张潇涵、朱利朋、戴文嘉、刘博伦、叶波、张增鑫。硕士研究生：朱颖文、肖琳、肖晔、梁静、郭逸文。指导教师：陈薇、白颖。硕士研究生：肖晔。指导教师：陈薇、是霏、史文娟。

四、插页图片底图来源：

庭园：[清]袁起 随园图 局部 南京博物院藏

私园：[明]文衡山拙政园诗画册 上 小飞虹 中华书局玻璃版部制版

景园：[明]沈周 东庄图册 局部 南京博物院藏

Editing Notes

1. Maps and Drawings

The maps and drawings of this book stem from the surveying and mapping collection of traditional landscape compiled by Southeast University (the formerly Nanjing Institute of Technology) School of Architecture during the period from the 1950s till today. Previous publications include *Suzhou gudian yuanlin* (The Classical Gardens of Suzhou) by LIU Dunzhen, published by China Architecture and Building Press in October 1979, *Jiangnan lijing yishu* (Jiangnan Garden Arts) by PAN Guxi, published by Southeast University Press in April 2001, and *Jiangnan yuanlin tulu—tingyuan,jingguan jianzhu* (Jiangnan Garden Drawings—Courtyard and Landscape Architecture) by LIU Xianjue and PAN Guxi, published by Southeast University Press in September 2007. The collection was enriched by maps and drawings produced during the 2007 master's degree program and the work of architecture and landscape undergraduates during 2015-2017.

2. Compilation

CHEN Wei is responsible for the composition and classification of the book, with SHI Fei and SHI Wenjuan working on the details of the gardens introduced in the book. CHEN Wei also took the assignments of proofreading and correction, with SHI Fei, SHI Wenjuan, ZHENG Hua, GAO Wenxi and XIE Fei working on editing specific maps and drawings. CHEN Wei is the final reviewer of the Chinese book content, while Alexandra Harrer, CHOU Yen Pang , and Jeff Zou are responsible for the English translation.

3. Major Events

1) *Suzhou gudian yuanlin* (The Classical Gardens of Suzhou)

In 1953, a team led by LIU Dunzhen with participation of the China Architecture Research Studio, jointly established by the Nanjing Institute of Technology and the East China Architecture Design and Research Institute, made a survey to the classical gardens in Suzhou. Later, a project, also led by Liu with participation of the architectural history faculties from the Nanjing Institute of Technology Department of Architecture and the former Ministry of Building Construction Research Institute, was launched to further investigate and map the major classical gardens in Suzhou. *Suzhou gudian yuanlin* was first published in 1960, then revised and enhanced in 1963.

Together with the professors and students from the former Nanjing Institute of Technology Department of Architecture, ZHU Mingquan, SHEN Guoyao, ZHAN Yongwei, JIN Qiyin, and FU Gaojie from the former Ministry of Building Construction Research Institute Nanjing Branch made major contributions to the mapping, drawing, and photographing. Furthermore, PAN Guxi, LIU Xianjue, LE Weizhong, GUO Husheng, YE Juhua, and LIU Xujie worked on the enhanced edition published in 1973.

2) *Jiangnan yuanlin tulu* (Jiangnan Garden Drawings)

June-July 1963: Undergraduate students (class of 1966) of the Nanjing Institute of Technology Department of Architecture surveyed and mapped the extant garden buildings in Suzhou, Wuxi, and Yangzhou under the tutorship of PAN Guxi, LIU Xianjue, DU Shunbao, CHEN Xiang, LI Boxian, and XU Dingliang.

December 1963: Undergraduate students (class 1966) of the Nanjing Institute of Technology Deptartment of Architecture survey and mapped the extnat courtyard and garden buildings in Nantong and Hangzhou, under the tutorship of, among others, PAN Guxi, LIU Xianjue, LIU Xianjue, DU Shunbao, and CHEN Xiang.

July 1964: Some 1966' undergraduate students and faculty members of the Nanjing Institute of Technology Department of Architecture further worked on the maps and drawings under the tutorship of PAN Guxi.

3) Supplements to *Jiangnan lijing yishu* (Jiangnan Garden Arts)

May 1981: PAN Guxi, DU Shunbao, and HE Jianzhong worked on the layout and section plans of Jichang Garden in Wuxi.

May 1982: A design team made up of the 1982' undergraduate students from the Nanjing Institute of Technology Deptartment of Architecture mapped the extant gardens in Xishan, Dongting, under the tutorship of PAN Guxi.

June 1982: The 1983' undergraduates of the Nanjing Institute of Technology Department of Architecture mapped the extant gardens in Yangzhou, Jiaxing, and Hangzhou, under the tutorship of, among others, DU Shunbao, ZHUANG Jinyuan, and ZHENG Guangfu.

1996: LI Zhen, ZHA Qun, and ZHANG Zhongmin of Southeast University mapped the Qi Garden and Yanyulou, under the tutorship of ZHU Guangya. The layout map of Qi Garden was redrawn based on the original provided by the Zhejiang Provincial Institute of Cultural Heritage and Archeology.

August 1996: The 1997' undergraduates of the Nanjing Institute of Technology Deptartment of Architecture mapped some extant garden buildings in Nanjing and Taizhou, under the tutorship of DING Hongwei, CHEN Wei, and LI Haiqing.

June 1997: GUO Huayu and ZHA Qun of Southeast University mapped Tuisi Garden.

September 1997: The 1983' undergraduates at the Nanjing Institute of Technology Deptartment of Architecture surveyed and mapped again Zhan Garden, under the tutorship of CHENG Yuning.

September 1997: HU Shi and FENG Wei surveyed and mapped again Guo's Residence in Hangzhou.

4) Mapping Efforts in Recent Years

November 2007: The master's degree students YANG Jun, XUE Xuan, and QI Zhao mapped Qiao Garden in Taizhou, under the tutorship of CHEN Wei.

August-September 2015: Together with the master's degree students YE Zhi, GUO Yiwen, MENG Yang, and GAO Wenjuan, a group of 2012' undergraduates including ZHANG Yue, QIAN Zhida, SHAO Jiandong, XU Wenting, LIU Shuo, HUANG Yuxing, YANG Yiran, ZUO Yanan, XIE Wenhui, XIA Yi, SHENG Suchen, SONG Jia, FENG Yuan, ZHU Mennan, and YAN Heng mapped the Garden of Cultivation in Suzhou, under the tutorship of CHEN Wei, JIA Tingli, and SHI Fei.

August-September 2016: Together with the master's degree students ZHU Yingwen, XIAO Ye, YU Hao, and LIANG Jing, a team of 2013' undergraduates including HUA Zhengchen, YU Guangzhou, YANG Yiming, ZHANG Ya, WU Jiangyuan, MA Liling, MA Lin, CHEN Shuang, DAI Wenyi, TAO Boshuang, SHI Junjie, GAO Xue, CUI Jilu, LI Hao, and LIU Xiaowen mapped the Couple's Garden Retreat in Suzhou, under the tutorship of CHEN Wei and BAI Ying.

August-September 2017: Together with the master's degree student XIAO Ye, a team of 2014' undergraduates including YANG Fan, XU Wen, GUO Jiaqi, GAO Wenxi, HUAI Yuhan, SHANG Ningyue, ZHENG Hua, LU Yizhu, CHEN Hao, ZHANG Xiaohan, ZHU Lipeng, DAI Wenjia, LIU Bolun, YE Bo, and ZHANG Zengxin mapped Canglangting in Suzhou, under the tutorship of CHEN Wei, SHI Fei, and SHI Wenjuan.

4. Sources of pictures on interleaves

Courtyard Gardens: Part of Sui Yuan Garden Map by YUAN Qi in Qing Dynasty, Collection of Nanjing Museum
Private Gardens: Xiaofeihong in WEN Hengshan Humble Administrator's Garden Poems and Pictures Album in Ming Dynasty, Making by the Glass Plate Department of Zhonghua Book Company
Landscape Parks: Part of Dongzhuang Atlas by SHEN Zhou in Ming Dynasty, Collection of Nanjing Museum

图书在版编目（CIP）数据

江南园林 =JIANGNAN GARDENS：汉英对照 / 陈薇，是霏主编；东南大学建筑学院编写 . —北京：中国建筑工业出版社，2019.12

（中国古建筑测绘大系·园林建筑）

ISBN 978-7-112-24562-8

Ⅰ . ①江… Ⅱ . ①陈… ②是… ③东… Ⅲ . ①古典园林—建筑艺术—华东地区—图集 Ⅳ. ① TU-092.2

中国版本图书馆CIP数据核字（2019）第286228号

丛书策划 / 王莉慧

责任编辑 / 李　鸽　刘　川

英文审稿 / ［奥］荷雅丽（Alexandra Harrer）

责任设计 / 付金红

责任校对 / 王　烨

中国古建筑测绘大系·园林建筑

江南园林

东南大学建筑学院　编写

陈　薇　是　霏　主编

Traditional Chinese Architecture Surveying and Mapping Series: Garden Architecture

JIANGNAN GARDENS

Compiled by School of Architecture, Southeast University

Edited by CHEN Wei, SHI Fei

*

中国建筑工业出版社出版、发行（北京海淀三里河路9号）

各地新华书店、建筑书店经销

北京方舟正佳图文设计有限公司制版

北京雅昌艺术印刷有限公司印刷

*

开本：787毫米×1092毫米　横1/8　印张：47½　字数：1219千字

2021年10月第一版　2021年10月第一次印刷

定价：**358.00**元

ISBN 978-7-112-24562-8

（35218）